Springer Theses

Recognizing Outstanding Ph.D. Research

Aims and Scope

The series "Springer Theses" brings together a selection of the very best Ph.D. theses from around the world and across the physical sciences. Nominated and endorsed by two recognized specialists, each published volume has been selected for its scientific excellence and the high impact of its contents for the pertinent field of research. For greater accessibility to non-specialists, the published versions include an extended introduction, as well as a foreword by the student's supervisor explaining the special relevance of the work for the field. As a whole, the series will provide a valuable resource both for newcomers to the research fields described, and for other scientists seeking detailed background information on special questions. Finally, it provides an accredited documentation of the valuable contributions made by today's younger generation of scientists.

Theses may be nominated for publication in this series by heads of department at internationally leading universities or institutes and should fulfill all of the following criteria

- They must be written in good English.
- The topic should fall within the confines of Chemistry, Physics, Earth Sciences, Engineering and related interdisciplinary fields such as Materials, Nanoscience, Chemical Engineering, Complex Systems and Biophysics.
- The work reported in the thesis must represent a significant scientific advance.
- If the thesis includes previously published material, permission to reproduce this must be gained from the respective copyright holder (a maximum 30% of the thesis should be a verbatim reproduction from the author's previous publications).
- They must have been examined and passed during the 12 months prior to nomination.
- Each thesis should include a foreword by the supervisor outlining the significance of its content.
- The theses should have a clearly defined structure including an introduction accessible to new PhD students and scientists not expert in the relevant field.

Indexed by zbMATH.

More information about this series at https://link.springer.com/bookseries/8790

Gabin Yoon

Theoretical Study on Graphite and Lithium Metal as Anode Materials for Next-Generation Rechargeable Batteries

Doctoral Thesis accepted by
Seoul National University, Seoul, South Korea

Springer

Author
Dr. Gabin Yoon
Department of Materials Science
and Engineering
Seoul National University
Seoul, Korea (Republic of)

Supervisor
Prof. Kisuk Kang
Department of Materials Science
and Engineering
Seoul National University
Seoul, Korea (Republic of)

ISSN 2190-5053 ISSN 2190-5061 (electronic)
Springer Theses
ISBN 978-981-13-8913-9 ISBN 978-981-13-8914-6 (eBook)
https://doi.org/10.1007/978-981-13-8914-6

This Springer imprint is published by the registered company Springer Nature Singapore Pte Ltd.
The registered company address is: 152 Beach Road, #21-01/04 Gateway East, Singapore 189721, Singapore

Supervisor's Foreword

Rechargeable batteries have enriched our lives by allowing us to ubiquitously use electric energy in various devices from cellphones to electric vehicles. To extend the success of the rechargeable batteries, the development of new technologies beyond the current Li-ion batteries is of great importance to meet the various demands for energy storage. In this regard, tremendous efforts have been made by research groups including ours, and as a continuing contribution, this thesis describes theoretical studies on the electrode materials for rechargeable batteries, specifically on the graphite and the lithium metal during the six years of doctoral study of Gabin Yoon at Seoul National University.

The first two parts of the thesis deal with the Na intercalation chemistry in graphite, which has been regarded impossible until recent years. The origin of unstable Na-ion intercalation in graphite is systematically investigated, and an innovative strategy of Na intercalation via solvent co-intercalation is introduced to utilize graphite as an anode for Na-ion batteries. Furthermore, conditions of solvents for reversible co-intercalation are proposed as a result of the combination of theoretical efforts and experimental observations. These findings offer an insight on the general intercalation-based electrochemical systems including guest ions, solvents, and an intercalation host, and hint at a strategy to tailor the intercalation behavior for various guest ions. The last part of the thesis describes the deposition and stripping behavior of Li metal in electrochemical system. Li metal has been considered as a holy grail for an anode material in Li rechargeable batteries, while practical and fundamental bottlenecks have not been resolved yet. Continuum mechanics simulation with the various electrochemical parameters revealed the origin of dendritic Li growth as well as the factors determining the Li deposition behavior. More importantly, it is revealed that the relative stripping rate is a critical factor to the generation of Li metal debris, contrary to the conventional belief that the fast deposition/stripping causes the Li metal detachment. This thesis provides a fundamental understanding on the reaction mechanism of the most prominent anode materials, which opens up the wide opportunities for the design of electrode materials for next-generation rechargeable batteries.

Seoul, South Korea
April 2019

Prof. Kisuk Kang

Abstract

A worldwide drift for sustainable society has led to the intensive development of the electrochemical energy storage system (ESS). Among the various ESSs, Li-ion batteries (LIBs) have drawn widespread attention for various applications because of their high energy density, power capability, and stable cyclability. However, the energy density of LIBs is limited, and their production cost is unstable due to the uneven worldwide distribution and the limited supply of Li. In this regard, it is of great importance to search for electrodes for next-generation rechargeable batteries in order to meet an ever-growing variety of demands and improve the energy storage performances. In this thesis, I present theoretical investigation on anode materials for next-generation rechargeable batteries, particularly on the graphite for Na-ion batteries and the metallic Li for Li metal batteries.

In Chap. 2, the mechanism of solvated-Na-ion intercalation in graphite is investigated using *operando* X-ray diffraction analysis, electrochemical titration, real-time optical observation, and density functional theory (DFT) calculations. The ultra-fast intercalation is demonstrated in real time using millimeter-sized highly ordered pyrolytic graphite, in which instantaneous insertion of solvated-Na-ions occurs (in less than 2 s). The formation of various stagings with solvated-Na-ions in graphite is observed and precisely quantified for the first time. The atomistic configuration of the solvated-Na-ions in graphite is proposed based on the experimental results and DFT calculations. The correlation between the properties of various solvents and the Na-ion co-intercalation further suggests a strategy to tune the electrochemical performance of graphite electrodes in Na rechargeable batteries.

In Chap. 3, through a systematic investigation of a series of alkali metal (AM) graphite intercalation compounds (AM-GICs, AM = Li, Na, K, Rb, Cs) in various solvent environments, I demonstrate that the unfavorable local Na–graphene interaction primarily leads to the instability of Na-GIC formation but can be effectively modulated by screening Na-ions with solvent molecules. Moreover, it is shown that the reversible Na intercalation into graphite is possible only for specific conditions of electrolytes with respect to the Na–solvent solvation energy and the lowest unoccupied molecular orbital level of the complexes. I believe that these conditions are applicable to other electrochemical systems involving guest ions and an intercalation host and hint at a general strategy to tailor the electrochemical intercalation between pure guest ion intercalation and co-intercalation.

In Chap. 4, various aspects of the electrochemical deposition and stripping of Li metal are investigated with consideration of the reaction rate/current density, electrode morphology, and solid electrolyte interphase (SEI) layer properties to understand the conditions inducing abnormal Li growth. It is demonstrated that the irregular (*i.e.,* filamentary or dendritic) growth of Li metal mostly originates from local perturbation of the surface current density, which stems from surface irregularities arising from the morphology, defective nature of the SEI, and relative asymmetry in the deposition/stripping rates. Importantly, I find that the use of a stripping rate of Li metal that is slower than the deposition rate seriously aggravates the formation of disconnected Li debris from the irregularly grown Li metal. This finding challenges the conventional belief that high-rate stripping/plating of Li in an electrochemical cell generally results in more rapid cell failure because of the faster growth of Li metal dendrites.

Keywords Energy storage · First-principles calculation · Continuum mechanics · Batteries · Li metal anode · Graphite

Parts of this thesis have been published in the following documents:

1. H. Kim†, J. Hong†, G. Yoon†, H. Kim, K.-Y. Park, M.-S. Park, W.-S. Yoon, K. Kang, Sodium intercalation chemistry in graphite, *Energy Environ. Sci.* **2015**, 8, 2963
2. G. Yoon, H. Kim, I. Park, K. Kang, Conditions for reversible Na intercalation in graphite: Theoretical studies on the interplay among guest ions, solvent, and graphite host, *Adv. Energy Mater.* **2017**, 7, 1601519
3. G. Yoon, S. Moon, G. Ceder, K, Kang, Deposition and stripping behavior of lithium metal in electrochemical systems: Continuum mechanics study, *Chem. Mater.* **2018**, 30, 6769

(† indicates equal contributions)

Acknowledgements

First, I would like to express my utmost gratitude to my advisor, Prof. Kisuk Kang, for all the support, guidance, and inspiration during the six years of my graduate study. He allowed me to take a lot of opportunities and have various experience which helped a lot, and his enthusiasm always motivated me.

I am deeply grateful to Dong-Hwa Seo for taking so much time to teach me in my first year at the lab and giving a lot of help during my half-year stay in Berkeley. I specially thank Inchul Park and Byungju Lee for guidance and advices when I was struggling in my early years. My gratitude also goes to all other members in our theory team, Donghee, Do-Hoon, Kyungbae, and Byunghoon. I have been lucky to have a chance to work with such great colleagues. I extend my gratitude to Jongsoon Kim, Haegyeom Kim, Jihyun Hong, and Hyungsub Kim. Working with them was always joyful and I learned a lot from their insight. I would also like to acknowledge the rest of Advanced Energy Materials Lab members for all the inspirations. My greatest gratitude of course goes to my parents, who have always supported me since I was a little boy.

Finally, thanks to all of you who directly or indirectly helped me finish this work.

Contents

Chapter 1
Introduction

1.1 Demands for Energy Storage System

The soaring demand for energy along with the exhaustion of fossil fuels and their environmentally harmful outcome have led to the growing interest in the sustainable energy source, including solar, wind and hydroelectric power generation. These eco-friendly energy resources necessarily require an energy storage system (ESS), since the energy production highly depends on the environmental condition. In addition, as the portable electronic devices from cellphones to electric vehicles proliferate, the development of ESS becomes indispensable. For a last few decades, Li-ion batteries (LIBs) which stores chemical energy using reversible Li (de)intercalation to the electrode materials have drawn widespread attention for their high energy density, power capability and stable cyclability [1–3]. However, the performance improvement of LIBs is slow in progress, whereas the demands for high-performance ESSs are soaring as well as diversifying. For instance, applications such as mobile devices and electric vehicles require high energy density, while renewable energy plants need a large-scale and a low-cost ESS. In this regard, the development of next-generation rechargeable batteries is of importance in order to meet various demands.

1.2 Li-Ion Batteries

LIBs are representative energy storage system, which operates by the reversible conversion between chemical energy and electric energy. Because of their high energy density, power density and cycle stability, the application of LIBs ranges from small portable devices to large electric vehicles. A simple configuration of a cell comprising LIBs is illustrated in Fig. 1.1. Each cell consists of two electrodes (anode and cathode) where redox reactions occur, separated by electrolyte where Li

G. Yoon, *Theoretical Study on Graphite and Lithium Metal as Anode Materials for Next-Generation Rechargeable Batteries*, Springer Theses,
https://doi.org/10.1007/978-981-13-8914-6_1

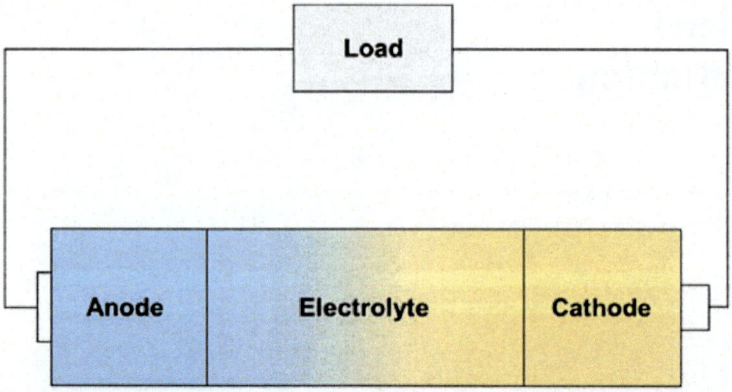

Fig. 1.1 Schematic picture of a cell configuration

ions flow and external circuit where electrons flow. During discharge, where chemical energy is converted to electrical energy, Li ions migrate from the anode to the cathode through the electrolyte, and electrons move through the electrical circuit. The driving force of this process is the difference of Li chemical potential in the cathode and the anode, which is voltage of a cell. Charge is a reverse process of discharge. Li ions migrate from the cathode to the anode, and the electrical energy is converted to chemical energy.

Performance of a cell is typically assessed by charge–discharge curve, illustrated in Fig. 1.2. Vertical axis of the curve shows the voltage, and horizontal axis of the curve describes the capacity, which is the amount of charge stored (extracted) during the charge (discharge) of a cell. Product of voltage and capacity per unit mass or volume of a cell is energy density. Therefore, in order to develop a battery with higher performance, it is important to discover electrodes with higher capacity while lowering the potential of anodes and the raising the potential of cathodes. Meanwhile,

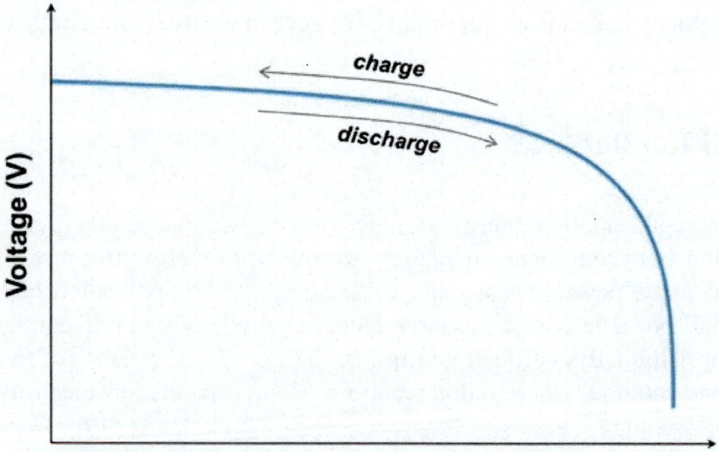

Fig. 1.2 Charge–discharge curve of a typical battery cell

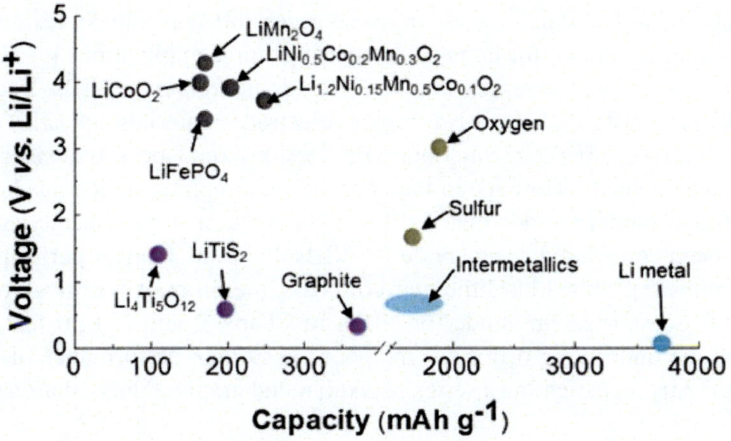

Fig. 1.3 Voltage and capacity of representative electrode materials for LIBs

the cyclic stability, the cost and the safety of a cell should also be significantly considered.

Figure 1.3 depicts the potential and the gravimetric capacity of representative electrode materials for LIBs. Among a variety of materials, intercalation-based materials are typically used as electrodes for LIBs, since the first commercialized LIB comprising $LiCoO_2$ cathode and graphite anode. This is because they serve as hosts for (de)intercalating Li ions, which ensures the structure stability during the cell operation, providing good cycle stability. However, the weight and the volume of the host framework structure limits the energy density of intercalation-based electrodes. In this regard, electrodes utilizing conversion or alloying mechanism have been intensively researched for post LIB system.

1.3 Post Li-Ion Batteries

As discussed in Sect. 1.1, a variety of next-generation rechargeable batteries has drawn extensive attention to meet the demands for future applications. In this book, two categories of post LIB systems are described in detail: Na-ion batteries for cost-effect and large-scale applications and Li metal batteries for high-energy-density systems.

1.3.1 Na-Ion Batteries

One of the promising candidates for post-LIBs is Na-ion batteries (NIBs), due to their similar electrochemistry to the established LIBs. In addition, the low cost of

Na compared to Li, which comes from the accessibility of the Na resources, makes NIB a strong candidate for large-scale energy storage applications [4–10].

Figure 1.4 shows the representative electrode materials for NIBs. Because of the chemical similarity of Li and Na, typical electrode materials for LIBs can be also utilized for NIBs, if their Na-containing version could be synthesized. However, there is one critical difference in the anode side. Graphite, which has been broadly utilized as a standard anode material for LIBs due to its high capacity and low cost, has not been considered as an anode for NIBs due to its inferior performance of Na storage [11–13]. This dilutes the merit of NIBs, therefore, it is critical to find a way to exploit graphite as an anode for NIBs. In Chaps. 2 and 3, I present the studies to reveal the underlying origin for the poor Na storage performance of graphite by systematically investigating a series of alkali metal graphite intercalation compounds

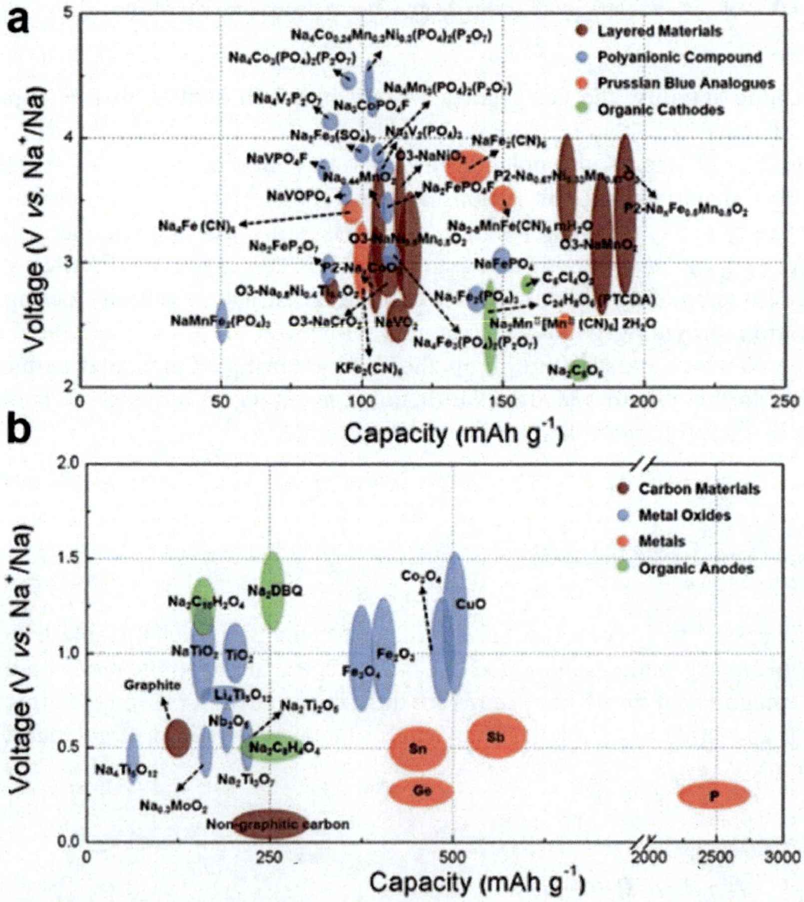

Fig. 1.4 Representative **a** cathode and **b** anode materials for Na-ion batteries. Reproduced with permission [5] from Wiley-VCH

(AM-GICs, AM = Li, Na, K, Rb, Cs) [14, 15]. In addition, I demonstrate the Na-solvent co-intercalation is an effective strategy for reversible Na intercalation into graphite. Furthermore, with the aid of the experimental measurements, the mechanism of solvated-Na-ion intercalation in graphite is investigated. The formation of various stagings of solvated-Na-ions in graphite is observed and precisely quantified, and the atomic arrangement of the solvated-Na-ions in graphite is proposed based on experiments and DFT calculations. Finally, I demonstrate that the reversible intercalation of solvated-Na-ions in graphite is possible for specific solvents, which satisfy the conditions regarding the Na-solvent solvation energy and the lowest unoccupied molecular level of solvated-Na-ion complexes.

1.3.2 Li Metal Batteries

The use of Li metal as an anode for next-generation batteries is also an effective approach in terms of an energy density improvement. While the conventional intercalation-based anodes show limited gravimetric capacity (~370 mAh g^{-1}) due to the weight of the intercalation host, Li metal anode has a potential of delivering the largest capacity (~3860 mAh g^{-1}) and the lowest redox potential (−3.04 V vs. standard hydrogen electrode), since it utilizes the reversible electrochemical plating/stripping of Li metal instead of the intercalation [16]. However, the safety issues arising from the dendritic growth of Li prevents the practical application. Various physical and chemical approaches have been applied to address this issue, however, the stable cycle performance under practical operation conditions is not yet achieved [17–29].

In Chap. 4, I present the fundamental investigation of Li electrodeposition and stripping [30]. Key factors affecting the electrochemical Li deposition/stripping behavior, such as the reaction rate, shape of the electrode surface, conductivity and uniformity of the SEI layer, and the effect of current density history were thoroughly controlled in continuum mechanics simulations to monitor their effect on the Li morphology evolution.

References

1. J.M. Tarascon, M. Armand, Issues and challenges facing rechargeable lithium batteries. Nature **414**, 359 (2001)
2. M. Armand, J.M. Tarascon, Building better batteries. Nature **451**, 652–657 (2008)
3. K. Kang, Y.S. Meng, J. Breger, C.P. Grey, G. Ceder, Electrodes with high power and high capacity for rechargeable lithium batteries. Science **311**, 977–980 (2006)
4. S.W. Kim, D.H. Seo, X.H. Ma, G. Ceder, K. Kang, Electrode materials for rechargeable sodium-ion batteries: potential alternatives to current lithium-ion batteries. Adv. Energy Mater. **2**, 710–721 (2012)
5. H. Kim, H. Kim, Z. Ding, M.H. Lee, K. Lim, G. Yoon, K. Kang, Recent progress in electrode materials for sodium-ion batteries. Adv. Energy Mater. **6**, 1600943 (2016)

6. P. Barpanda, T. Ye, S. Nishimura, S.C. Chung, Y. Yamada, M. Okubo, H.S. Zhou, A. Yamada, Sodium iron pyrophosphate: a novel 3.0 V iron-based cathode for sodium-ion batteries. Electrochem. Commun. **24**, 116–119 (2012)
7. S. Komaba, N. Yabuuchi, T. Nakayama, A. Ogata, T. Ishikawa, I. Nakai, Study on the reversible electrode reaction of $Na_{1-x}Ni_{0.5}Mn_{0.5}O_2$ for a rechargeable sodium-ion battery. Inorg. Chem. **51**, 6211–6220 (2012)
8. N. Yabuuchi, M. Kajiyama, J. Iwatate, H. Nishikawa, S. Hitomi, R. Okuyama, R. Usui, Y. Yamada, S. Komaba, P2-type $Na_x[Fe_{1/2}Mn_{1/2}]O_2$ made from earth-abundant elements for rechargeable Na batteries. Nat. Mater. **11**, 512–517 (2012)
9. H. Kim, G. Yoon, I. Park, K.Y. Park, B. Lee, J. Kim, Y.U. Park, S.K. Jung, H.D. Lim, D. Ahn, S. Lee, K. Kang, Anomalous Jahn–Teller behavior in a manganese-based mixed-phosphate cathode for sodium ion batteries. Energy Environ. Sci. **8**, 3325–3335 (2015)
10. J. Song, L. Wang, Y. Lu, J. Liu, B. Guo, P. Xiao, J.-J. Lee, X.-Q. Yang, G. Henkelman, J.B. Goodenough, Removal of interstitial H_2O in hexacyanometallates for a superior cathode of a sodium-ion battery. J. Am. Chem. Soc. **137**, 2658–2664 (2015)
11. D. Kundu, E. Talaie, V. Duffort, L.F. Nazar, The emerging chemistry of sodium ion batteries for electrochemical energy storage. Angew. Chem. Int. Ed. **54**, 3431–3448 (2015)
12. N. Yabuuchi, K. Kubota, M. Dahbi, S. Komaba, Research development on sodium-ion batteries. Chem. Rev. **114**, 11636–11682 (2014)
13. M.D. Slater, D. Kim, E. Lee, C.S. Johnson, Sodium-ion batteries. Adv. Funct. Mater. **23**, 947–958 (2013)
14. G. Yoon, H. Kim, I. Park, K. Kang, Conditions for reversible Na intercalation in graphite: theoretical studies on the interplay among guest ions, solvent, and graphite host. Adv. Energy Mater. **7**, 1601519 (2017)
15. H. Kim, J. Hong, G. Yoon, H. Kim, K.-Y. Park, M.-S. Park, W.-S. Yoon, K. Kang, Sodium intercalation chemistry in graphite. Energy Environ. Sci. **8**, 2963–2969 (2015)
16. J.W. Choi, D. Aurbach, Promise and reality of post-lithium-ion batteries with high energy densities. Nat. Rev. Mater. **1**, 16013 (2016)
17. C. Monroe, J. Newman, The impact of elastic deformation on deposition kinetics at lithium/polymer interfaces. J. Electrochem. Soc. **152**, A396–A404 (2005)
18. X. Chen, T.-Z. Hou, B. Li, C. Yan, L. Zhu, C. Guan, X.-B. Cheng, H.-J. Peng, J.-Q. Huang, Q. Zhang, Towards stable lithium-sulfur batteries: mechanistic insights into electrolyte decomposition on lithium metal anode. Energy Storage Mater. **8**, 194–201 (2017)
19. C. Zu, N. Azimi, Z. Zhang, A. Manthiram, Insight into lithium-metal anodes in lithium-sulfur batteries with a fluorinated ether electrolyte. J. Mater. Chem. A **3**, 14864–14870 (2015)
20. K. Yan, H.-W. Lee, T. Gao, G. Zheng, H. Yao, H. Wang, Z. Lu, Y. Zhou, Z. Liang, Z. Liu, S. Chu, Y. Cui, Ultrathin two-dimensional atomic crystals as stable interfacial layer for improvement of lithium metal anode. Nano Lett. **14**, 6016–6022 (2014)
21. E. Kazyak, K.N. Wood, N.P. Dasgupta, Improved cycle life and stability of lithium metal anodes through ultrathin atomic layer deposition surface treatments. Chem. Mater. **27**, 6457–6462 (2015)
22. A.C. Kozen, C.-F. Lin, A.J. Pearse, M.A. Schroeder, X. Han, L. Hu, S.-B. Lee, G.W. Rubloff, M. Noked, Next-generation lithium metal anode engineering via atomic layer deposition. ACS Nano **9**, 5884–5892 (2015)
23. G. Zheng, S.W. Lee, Z. Liang, H.-W. Lee, K. Yan, H. Yao, H. Wang, W. Li, S. Chu, Y. Cui, Interconnected hollow carbon nanospheres for stable lithium metal anodes. Nat. Nanotechnol. **9**, 618 (2014)
24. H. Lee, D.J. Lee, Y.-J. Kim, J.-K. Park, H.-T. Kim, A simple composite protective layer coating that enhances the cycling stability of lithium metal batteries. J. Power Sources **284**, 103–108 (2015)
25. N.W. Li, Y.X. Yin, C.P. Yang, Y.G. Guo, An artificial solid electrolyte interphase layer for stable lithium metal anodes. Adv. Mater. **28**, 1853–1858 (2016)
26. X. Wang, Y. Hou, Y. Zhu, Y. Wu, R. Holze, An aqueous rechargeable lithium battery using coated Li metal as anode. Sci. Rep. **3**, 1401 (2013)

27. Y. Lu, Z. Tu, L.A. Archer, Stable lithium electrodeposition in liquid and nanoporous solid electrolytes. Nat. Mater. **13**, 961 (2014)
28. Y. Lu, Z. Tu, J. Shu, L.A. Archer, Stable lithium electrodeposition in salt-reinforced electrolytes. J. Power Sources **279**, 413–418 (2015)
29. F. Ding, W. Xu, G.L. Graff, J. Zhang, M.L. Sushko, X. Chen, Y. Shao, M.H. Engelhard, Z. Nie, J. Xiao, X. Liu, P.V. Sushko, J. Liu, J.-G. Zhang, Dendrite-free lithium deposition via self-healing electrostatic shield mechanism. J. Am. Chem. Soc. **135**, 4450–4456 (2013)
30. G. Yoon, S. Moon, G. Ceder, K. Kang, Deposition and stripping behavior of lithium metal in electrochemical system: continuum mechanics study. Chem. Mater. **30**, 6769–6776 (2018)

Chapter 2
Na Intercalation Chemistry in Graphite

2.1 Introduction

Graphite is a universal host material which is capable of accommodating a wide range of guest species including alkali metals, alkaline earths, rare earths, halogens, protonic and Lewis acids between sp^2-bonded graphene layers [1]. Its versatility also allows multiple guest species to be intercalated simultaneously in the host structure through "co-intercalation" process. Graphite intercalation compounds (GICs), the resulted compounds of intercalation, manifest distinctive physical and chemical properties compared to pristine graphite. Their electronic/magnetic structures, optical properties and catalytic properties significantly vary with the concentration of the guest species as well as with the type of intercalants [2]. For instance, the electronic conductivity of GICs, especially along *c*-axis, dramatically increases with the concentration of ionized guest species [3–5]. The unique structure–property relationship of GICs enables graphite and GICs to be applied to diverse fields such as energy storage, chemical catalysts, electrical/thermal conductors, and so on [6–10].

One of the most important applications of GICs is energy storage. Graphite has been utilized as a standard anode material in Li ion batteries where it forms a series of binary GICs during intercalation and finally reaches the stoichiometry of LiC_6 [11, 12]. However, graphite has not been considered as an anode for Na-ion batteries, an alternative system for large-scale energy storage, due to the thermodynamic instability of the binary Na-intercalated GICs [13, 14]. As recent breakthrough, Jache et al. and our group facilitated Na ion storage in graphite by using solvated-Na ion intercalation, forming ternary GICs [15, 16]. However, the mechanism of solvated-Na ion intercalation in graphite is still rarely understood due to its complexity while

The content of this chapter has been published in *Energy & Environmental Science.* [Kim, H., Hong, J., Yoon, G. et al., *Energy & Environmental Science* **2015**, 8, 2963–2969.] Reproduced with permission from The Royal Society of Chemistry.

G. Yoon, *Theoretical Study on Graphite and Lithium Metal as Anode Materials for Next-Generation Rechargeable Batteries*, Springer Theses,
https://doi.org/10.1007/978-981-13-8914-6_2

the simple binary GIC system of Li-graphite has been almost completely investigated [17–20]. In this regard, it is important to clearly identify the structure and the configuration of ternary GICs considering the close relationship between the structure and the electrochemical properties such as redox potential and specific capacity, i.e., energy density, as well as cycle stability.

In our continuing effort to exploit the graphite as a promising anode in Na-ion batteries, herein we precisely investigate the crystal structures and stoichiometry of the ternary Na-ether-graphite system in the process of solvated-Na intercalation by using *operando* X-ray diffraction (XRD) analysis coupled with electrochemical titration. The formation of first-stage GIC was directly visualized by real-time measurement of graphite expansion. The density functional theory (DFT)-based calculations confirmed the parallel double stacking of solvated-Na ions between the graphite layers. Furthermore, by introducing various solvents, we provide insights on tuning energy storage potentials of solvated ion intercalation based electrodes.

2.2 Experimental and Computational Details

2.2.1 Materials

Natural graphite was purchased from Bay Carbon Inc. and used without any modification or post-treatment. Electrolytes were carefully prepared to maintain low H_2O content (<20 ppm). Na salt ($NaPF_6$) and molecular sieves were stored in a vacuum oven at 180 °C before use. Dried Na salts were dissolved in electrolyte solvents, including diethylene glycol dimethyl ether (DEGDME), tetraethylene glycol dimethyl ether (TEGDME), dimethoxyethane (DME) at 1 M. The solution was stirred at 80 °C for 2 days and molecular sieves were added to remove residual H_2O from the electrolyte solution.

2.2.2 Electrode Preparation and Electrochemical Measurements

Graphite electrodes were prepared by mixing the active material (natural graphite, 90 wt%) with polyvinylidene fluoride binder (PVDF, 10 wt%) in an N-methyl-2-pyrrolidone (NMP) solvent. The resulting slurry was uniformly pasted onto Cu foil, dried at 120 °C for 1 h and roll-pressed. The average electrode thickness and loading density were ~40 μm. Test cells were assembled in a glove box into a two-electrode configuration with a Na metal counter electrode. A separator of grade GF/F (Whatman, USA) was sonicated in acetone and dried at 120 °C before use. Electrochemical profiles were obtained over a voltage range of 2.5–0.001 V using a multichannel potentio-galvanostat (WonATech).

2.2.3 Operando *XRD Analysis*

For *operando* XRD analysis, the holed coin cells which was sealed by kapton film with epoxy were used. *Operando* XRD patterns for graphite were collected on the 5A XRD beamline at PLS-II. The wavelength of the X-ray beam was 0.7653 Å, and XRD patterns were recorded as a set of circles on a Mar 345-image plate detector in the transmission mode for about 1 min of exposure time. The storage ring was operated with electron energy of 3.0 GeV and a current 300 mA. The total recording time was about 2.6 min because of the scanning time of the image plate and transferring time of spectral information. The two theta angles of all the XRD patterns presented in this article have been recalculated to corresponding angles for $\lambda = 1.54$ Å, which is the wavelength of the conventional X-ray tube source with Cu Kα radiations, for easy comparison with other published results.

2.2.4 Computational Details

Geometric optimizations of $[\text{Na-ether}]^+$ complexes were conducted with the density functional theory (DFT) platform by using Gaussian 09 program [21]. All molecular structures were fully relaxed with B3LYP/6-311G level. Structural relaxations of stage 1 and stage 2 GICs were performed with Vienna ab initio simulation package (VASP) [22]. We used a projector-augmented wave (PAW) pseudopotential [23] as implemented in VASP, and exchange-correlation energy was dealt with Perdew–Burke–Ernzerhof (PBE) parameterized generalized gradient approximation (GGA) [24]. In addition, semi-empirical dispersion potential (DFT-D2) [25] was implemented to describe the van der Waals (vdW) interactions of graphite. All GIC structures were relaxed until the total energy of the system converges within 0.01 eV Å^{-1}.

2.3 Staging Behavior upon Na-Solvent Co-intercalation

To understand the structural evolution of graphite during the solvated-Na intercalation/deintercalation, we thoroughly collected synchrotron-based XRD patterns of graphite *operando* (5A XRD beamline at PLS-II). For the analysis, a coin-type Na half-cell with a pin-hole containing ~40-μm thick graphite electrode and ~100-μl of 1 M $NaPF_6$ in diethylene glycol dimethyl ether (DEGDME) electrolyte was discharged/charged under a current density of 20 mA $g^{-1}{}_{graphite}$. Each diffraction pattern was obtained within 2.6 min corresponding to the capacity of ~0.87 mAh $g^{-1}{}_{graphite}$, which enables the sensitive detection of the structural transformation along the minutest alteration of Na contents. Throughout the electrochemical sodiation and desodiation, the structure of the graphite electrode is reversibly

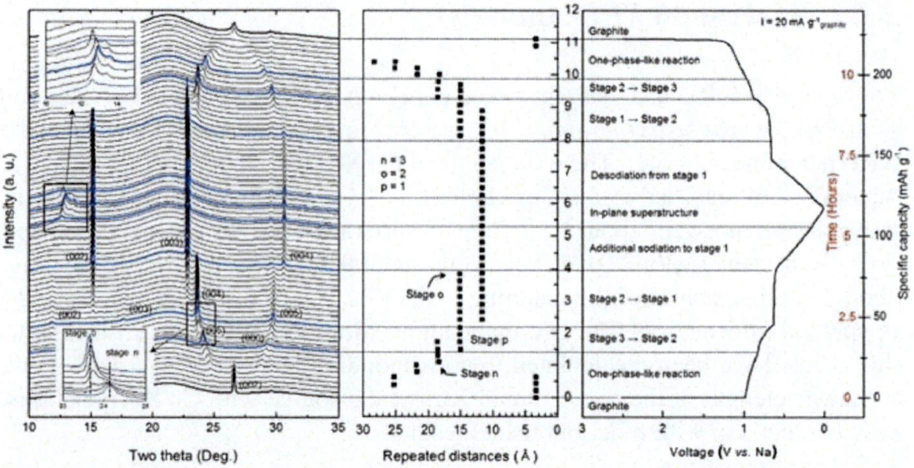

Fig. 2.1 *Operando* synchrotron X-ray diffraction analysis of the structural evolution of the ternary Na-ether-graphite system observed during electrochemical solvated-Na-ion intercalation and deintercalation into/out of graphite in Na | 1 M $NaPF_6$ in a DEGDME | graphite cell. Reproduced with permission from The Royal Society of Chemistry

transformed and restored (Fig. 2.1). We believe that this high reversibility of the structural transformation could lead to the stable capacity retention of graphite electrode over thousands of battery cycles [16] as previously reported. The details about this remarkable cyclability will be further discussed later.

The evolution of the XRD patterns of graphite exhibits a typical staging behavior during electrochemical discharge and charge. The (002) peak at 27° splits into two peaks which can be indexed as (00*l*) and (00*l* + 1), respectively [26]. Applying the Bragg's law to the graphite, we determined the *l*-value (see Eqs. 2.1 and 2.2).

$$d_{(00l)} = \frac{I_c}{l} \text{ and } d_{(00l+1)} = \frac{I_c}{l+1} \tag{2.1}$$

$$l = \frac{1}{\left[\frac{\sin\theta_{00l+1}}{\sin\theta_{00l}} - 1\right]} \tag{2.2}$$

where $d_{(00l)}$ and $d_{(00l+1)}$ are the d-spacing values of (00*l*) and (00*l* + 1) planes, respectively, and I_c is the *c* lattice parameter of each stage, corresponding to the repeated distance.

At the initial period of sodiation (discharge), the graphite transforms into GIC with stages changing sensitively to a small alteration of the Na content. The rapid stage transformation occurs until the specific capacity of graphite reaches 31 mAh g^{-1} (Na: C = 1/72), forming a stage *n* GIC. After a following biphasic reaction of forming stage *o* between 1.5 and 2.2 h of sodiation (Na: C = 1/50), a final stage *p* evolves. The pure phase of stage *p* GIC was observed after 4 h of sodiation within a wide range of Na/C from 1/28 to 1/21 with no significant change in XRD patterns except for the evolution of new peaks at 12–14° at a high concentration of solvated-Na ions in graphite as the state of discharge exceeds 91%. While further study is required for clarification, we expect that the new peaks originate from the in-plane superstructural ordering of the Na ions and ether solvent molecules, which will be discussed in detail later. Note that the *c* lattice parameters (I_c) of stage *n*, *o* and *p* GICs correspond to 18.50, 15.06 and 11.62 Å. The differences between I_c values are approximately 3.44 Å, similar to the interlayer distances of the pristine graphite (3.35 Å), which allows us to index the stage numbers *n*, *o* and *p* as n, $n-1$, and $n-2$, respectively. In addition, considering that 11.62 Å (I_c of stage *p*) is smaller than 13.4 Å (= 3.35 Å × 4), *p* should be less than 4.

To determine the final stage index, we measured the height expansion of graphite in real-time while intercalating Na and DEGDME solvent. We employed chip-type highly ordered pyrolytic graphite (HOPG) with 10 × 10 × 1.5 mm dimension (0.8° mosaic spread) to adjust the scale of alteration which enables the measurement with digital camcorder (Fig. 2.2a). The HOPG chip was placed on the Na metal foil and between the two metal cubes to make contact in both parallel and vertical directions to the graphite layers. The intercalation does not take place in the absence of the electrolyte although the HOPG and Na metal are physically connected (Fig. 2.2a, b). However, as soon as we drop the electrolyte at interface, the HOPG instantly and rapidly swells from the interface, indicative of the Na-ether co-intercalation. In a few minutes, under exposure to the excess electrolyte, the height of HOPG converges to 278% of the pristine state. But we found an imbalance of the height resulting from the loose contact between HOPG and Na metal described as yellow triangle in Fig. 2.2a (row 3, column 4). By tightening the contact, the expansion further proceeded which implies that the Na diffusion occurs along *a* and *b* axes. In the end, the height of HOPG was saturated at 346% of the pristine state, which is close to the theoretically estimated degree of height expansion (347%) in case of forming stage 1 GIC with *c* lattice parameter of 11.62 Å (Fig. 2.2c).

2.4 Na-Solvent Co-intercalation into Graphite Structure

For better understanding on the solvated-Na intercalation mechanism, we conducted a detailed investigation on how Na ions and DEGDME molecules are stored in graphite host. First, we clarified the amount of intercalated DEGDME molecules per Na ion in the graphite. The weight changes of graphite electrode were monitored at several states of charge and discharge (Fig. 2.3a). To exclude the weight change from the residual free solvents adsorbed on the electrode surface, the graphite electrodes were dried at 60 °C for 24 h after disassembled from the coin cells. All the sampling procedures were conducted in an Ar-filled glove box to prevent the contamination. The dashed lines show the theoretical weight changes of the graphite when different number (b-value) of DEGDME molecules per Na ion are intercalated into the graphite. It is clear that the weight of the graphite electrode follows the b = 1

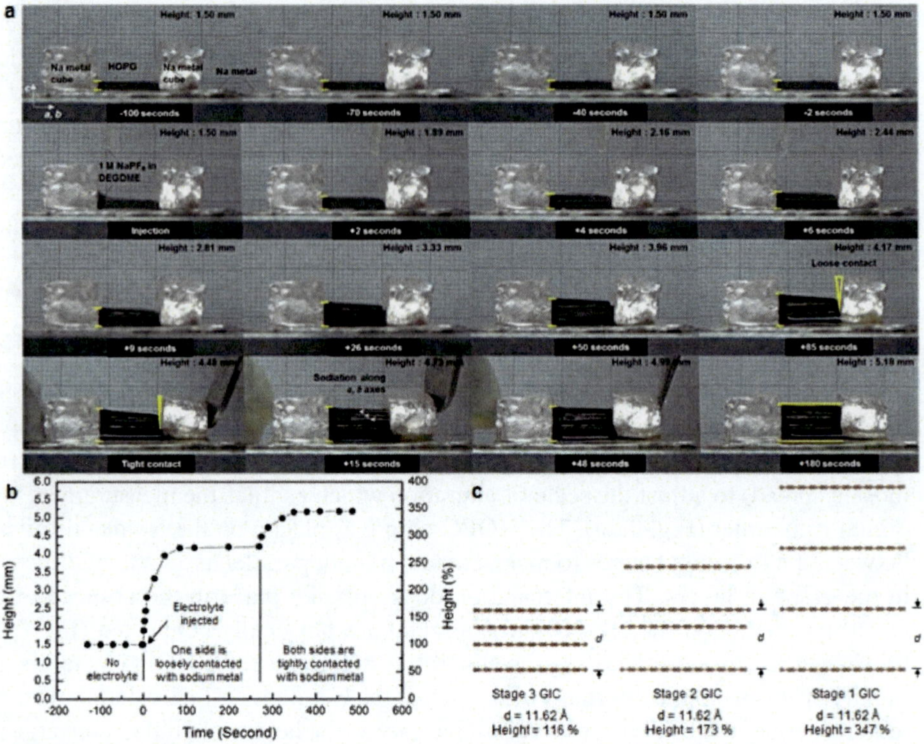

Fig. 2.2 Expansion of highly ordered pyrolytic graphite (HOPG) along the *c* axis with chemical Na-ether intercalation. **a** Real-time snapshots of HOPG under direct contact with Na metal before and after exposure to 1 M $NaPF_6$ in DEGDME solution, **b** height of HOPG measured during the intercalation process, and **c** schematic comparison of the height and *c* lattice parameter of graphite for hypothetical stage 3 GIC, 2 GIC, and stage 1 GIC. Reproduced with permission from The Royal Society of Chemistry

line, which corresponds to the 1:1 ratio of DEGDME molecule and Na ion in GICs, while the experimental values are slightly higher than the theoretical ones because of the solid-electrolyte-interphase (SEI) formation upon discharge. It is also supported by the EDS analysis (Fig. 2.3b) which gives that a ratio of O to Na atoms in GICs is approximately three, confirming that one DEGDME molecule is intercalated with one Na ion.

We further built model structures to verify how Na ions and DEGDME molecules are intercalated in the graphite electrode, given that one Na ion is intercalated into graphite with one DEGDME molecule. Since it is generally believed that cation and solvent are co-intercalated into graphite maintaining the solvation complex form [16, 27–29], we first identified the most stable structure of $[\text{Na-DEGDME}]^+$ complex by DFT calculations (Fig. 2.4), where three oxygen atoms of DEGDME are coordinated with Na ions. Among various possible configurations of stage 1 GICs, two representative models are chosen for comparison (Fig. 2.3c, d). One is a simple stage 1 GIC with single $[\text{Na-DEGDME}]^+$ complex intercalation in the middle of the graphite gallery

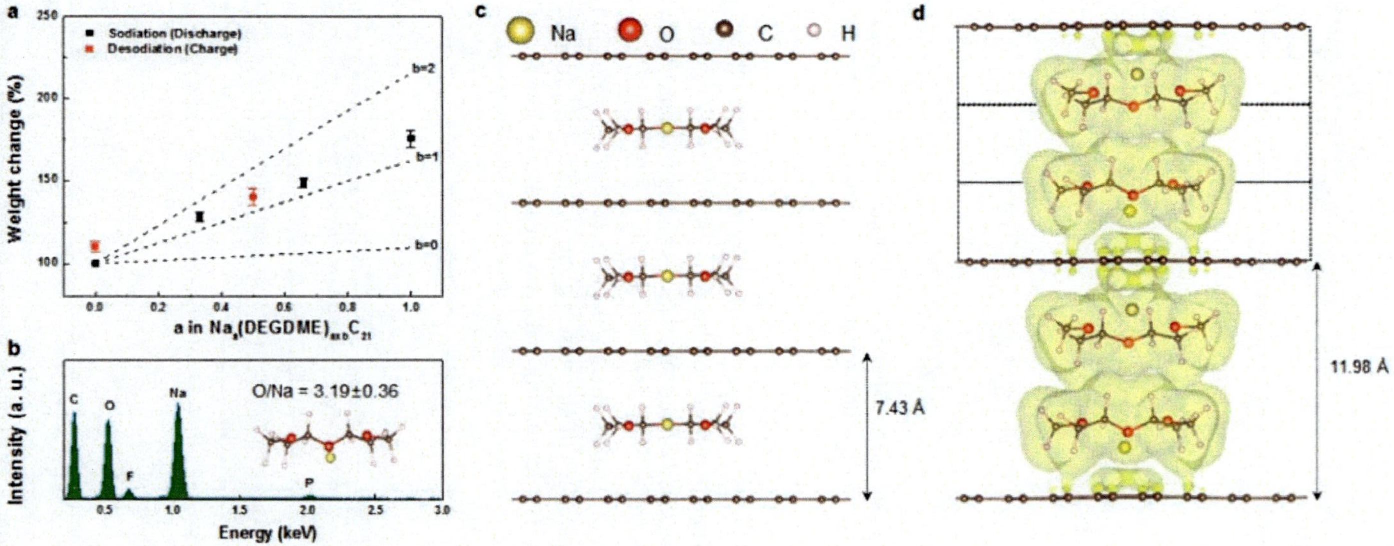

Fig. 2.3 Geometry of solvated-Na intercalated graphite. **a** Weight change of graphite measured at various state of charge and discharge. **b** Energy-dispersive X-ray spectroscopy analysis of discharged sample. Calculated structure when **c** one $[Na\text{-}DEGDME]^+$ complex is and **d** two $[Na\text{-}DEGDME]^+$ complexes are intercalated in the interlayer spacing in graphite. Reproduced with permission from The Royal Society of Chemistry

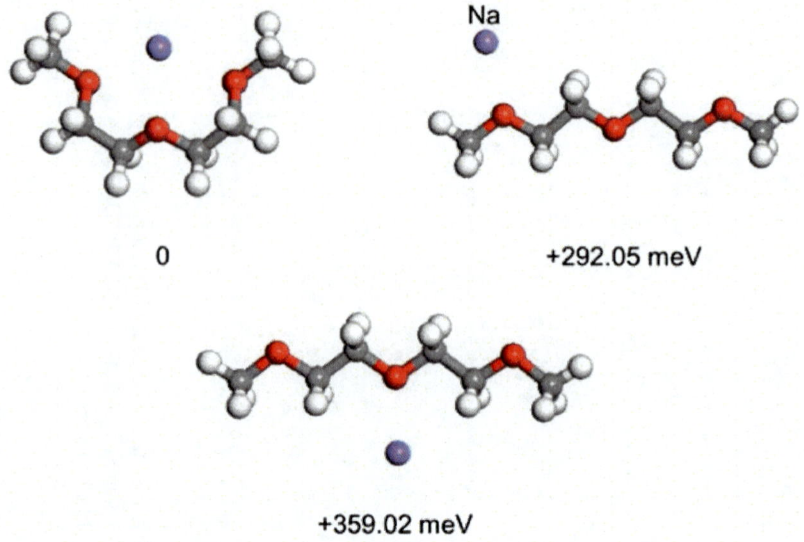

Fig. 2.4 Various configurations of [Na-DEGDME]$^+$ solvation complex. Energies relative to the most stable configuration are shown with each structure. Reproduced with permission from The Royal Society of Chemistry

(Fig. 2.3c). The other model possesses doubly stacked [Na-DEGDME]$^+$ complexes located at the one-third and two-third height of the graphite gallery which might not be intuitively accepted (Fig. 2.3d). Note that various configurations of [Na-DEGDME]$^+$ complex in graphite host were considered for doubly stacked complex intercalation in graphite (Fig. 2.5), and the most stable model structure is described in Fig. 2.3d. The calculated interlayer distances of single stack and double stack models were 7.43 Å and 11.98 Å, respectively. Comparing with the measured c lattice parameter of the

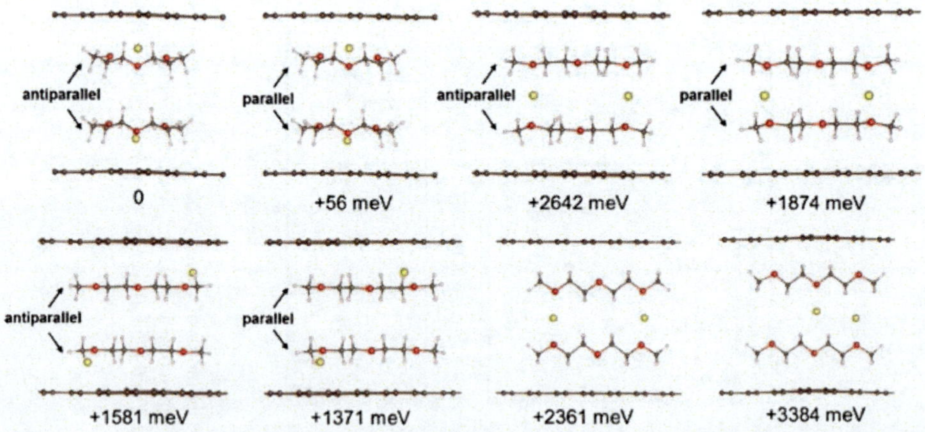

Fig. 2.5 Various configurations of double complex intercalation of [Na-DEGDME]$^+$ in stage 1 structure. Reproduced with permission from The Royal Society of Chemistry

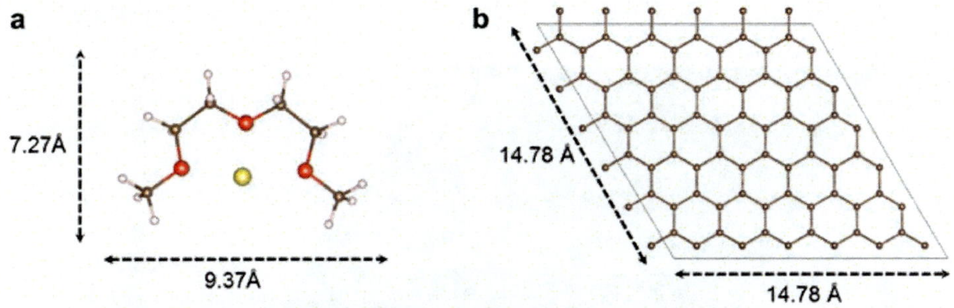

Fig. 2.6 Calculated planar area of **a** [Na-DEGDME]$^+$ and **b** graphene layer. For convenience, rectangular shape of [Na-DEGDME]$^+$ is assumed and intermolecular interactions are considered in Fig. 2.6a. The planar areas of [Na-DEGDME]$^+$ and graphene layer are 68.13 Å^2 per complex, and 2.64 Å^2 per carbon atom, respectively. Reproduced with permission from The Royal Society of Chemistry

stage 1 GIC (11.62 Å, see Fig. 2.1), the double stack model exhibited only 3.1% of difference, whereas the single stack model showed 36% of difference. This implies the plausibility of double stack model. To gain additional information for structure determination, we calculated the volume occupied by [Na-DEGDME]$^+$ complexes in graphite interlayers, and compared it with the amount of charge uptake in stage 1 GIC. Considering the planar area of [Na-DEGDME]$^+$ complex (68.13 Å per complex) and graphene layers (2.64 Å per carbon atom), the maximum Na concentration in the stage 1 GIC is [Na-DEGDME]$C_{25.8}$ and [Na-DEGDME]$C_{12.9}$ for single stack and double stack model, respectively, based on simple calculation (Fig. 2.6). The experimentally obtained reversible capacity of the graphite electrode is ~110 mAh g^{-1} which corresponds to one [Na-DEGDME]$^+$ complex per 21 carbon atoms. This value exceeds the maximum density of single [Na-DEGDME]$^+$ complex intercalation model, and it is below that of doubly stacked complex intercalation model, indicating the geometry of stage 1 GIC is more likely to have the double stack model shown in Fig. 2.3d.

The thermodynamic stability of the proposed structure (Fig. 2.3d) was investigated by DFT calculations to further support our model structure. We found that the reaction enthalpy of [Na-DEGDME]$^+$ co-intercalation into graphite is −0.87 eV, confirming that the co-intercalation of Na and DEGDME into graphite is feasible. We also confirmed that the formation of triple stack model is thermodynamically unstable, whereas the formation of double stack model is thermodynamically favorable. We note that the vibrational, configurational, rotational entropies of [Na-DEGDME]$^+$ complex can play an important role in the reaction Gibbs free energy. However, since [Na-DEGDME]$^+$ stays freely both before and after the insertion reaction, we believe that vibrational and rotational entropy change will have limited effect on reaction free energy. The contribution of configurational entropy is also negligible, considering that [Na-DEGDME]$^+$ intercalants densely occupy the interlayer space between graphene layers.

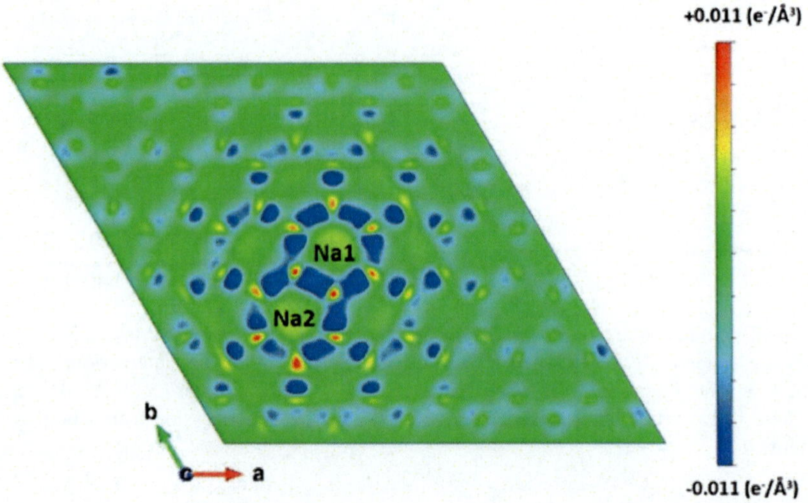

Fig. 2.7 In-plane charge distribution at graphene layer after the intercalation of [Na-DEGDME]$^+$ complex. Intercalated Na^+ ions donate electrons to graphene layer, forming an ionic bond with electrons from C–C bonds in graphene layer. Reproduced with permission from The Royal Society of Chemistry

We further looked into the charge interaction between the intercalated [Na-DEGDME]$^+$ complex and graphene layers. Yellow region in Fig. 2.3d indicates the charge gained after [Na-DEGDME]$^+$ intercalation (isosurface was set to 0.017 $e^-/Å^3$). A major portion of electron cloud, which comes from solvent molecules, resides in approximately one-third, two-thirds of unit cell height. Electrons from Na ions move towards graphene layers, forming ionic bonds with electrons donated from C–C bonds in graphene layers (Fig. 2.7), which is also observed in Li intercalated graphite [30]. Addition of charge density in graphene layers makes them push each other, resulting in a slight larger interlayer distance in GICs (3.44 Å) compared to pristine graphite (3.35 Å), which is in a good agreement with experimental observation in Fig. 2.1.

2.5 Solvent Dependency on Electrochemical Properties

To extend our knowledge of how the solvated Na storage properties of graphite are related to the solvent species, we conducted electrochemical measurements using three linear ether solvent species with different chain lengths (i.e. DME, DEGDME, and TEGDME). As shown in Fig. 2.8a, the specific capacity and shape of charge/discharge profiles were similar in all cases, indicating that Na storage would proceed through similar staging behavior, which is further supported by ex situ XRD analyses (Figs. 2.9, 2.10 and 2.11). We also confirmed that DME/Na and

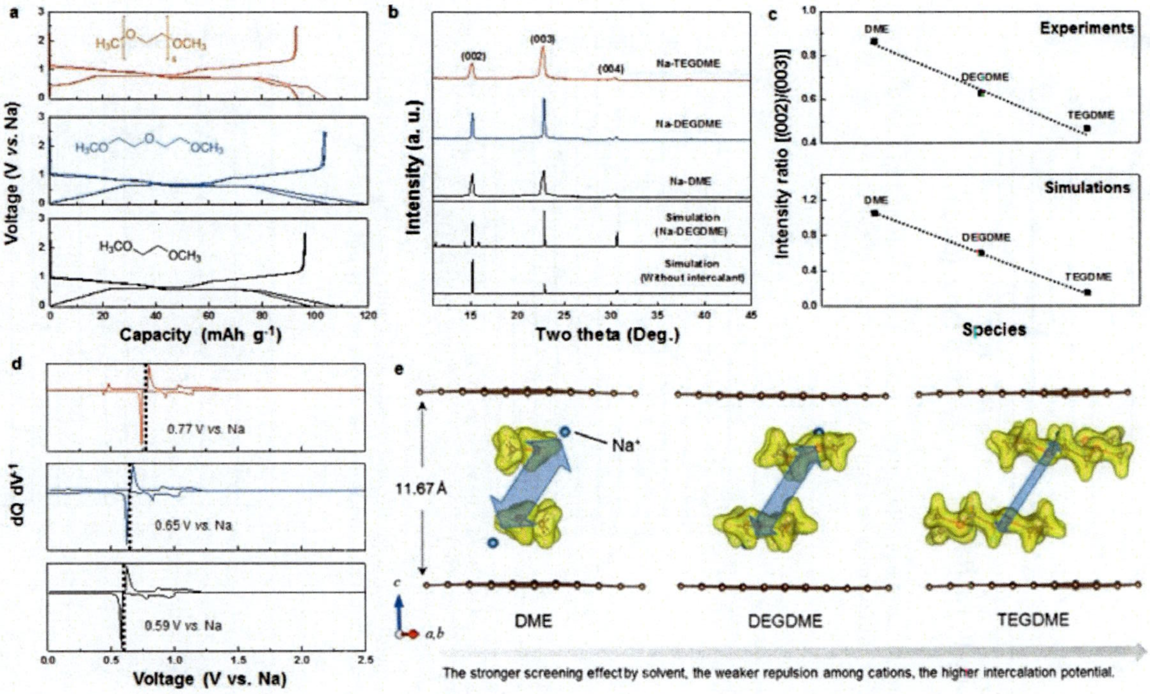

Fig. 2.8 Na storage potential depending on solvent species. **a** Typical charge/discharge profiles of graphite electrode with three electrolyte solvents (inset: solvent molecules) in Na half-cell configuration. **b** X-ray diffraction (XRD) patterns of solvated-Na intercalated graphite compounds and simulated XRD patterns of expanded graphite with repeated distance of 11.66 Å with/without intercalants of $[Na\text{-}DEGDME]^+$ complex. **c** Intensity ratio of (002)/(003) when three electrolyte solvents are intercalated into graphite with Na ions. **d** dQ dV^{-1} analysis of graphite electrode with three electrolyte solvents. **e** Schematic of the screening effect of the solvent on the Na storage potential. Reproduced with permission from The Royal Society of Chemistry

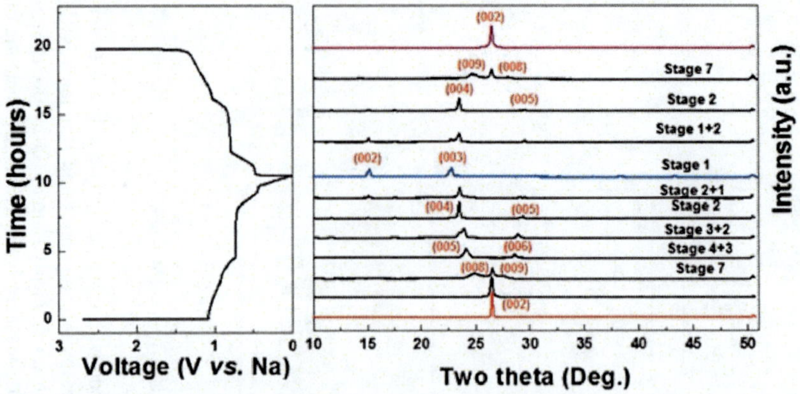

Fig. 2.9 Typical charge/discharge profile and ex-situ XRD patterns using 1 M $NaPF_6$ in TEGDME electrolyte. Reproduced with permission from The Royal Society of Chemistry

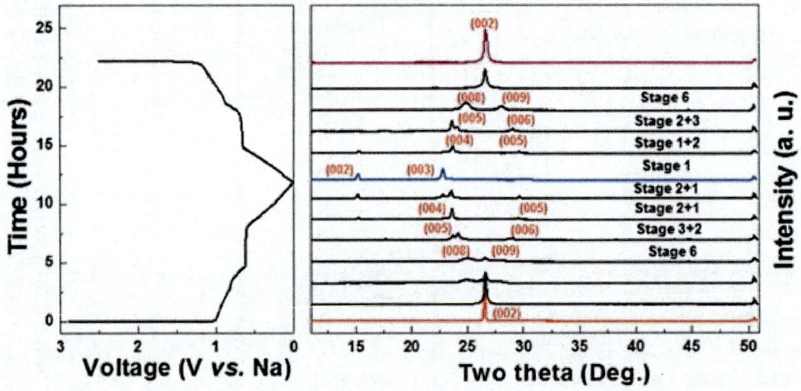

Fig. 2.10 Typical charge/discharge profile and ex-situ XRD patterns using 1 M $NaPF_6$ in DEGDME electrolyte. Reproduced with permission from The Royal Society of Chemistry

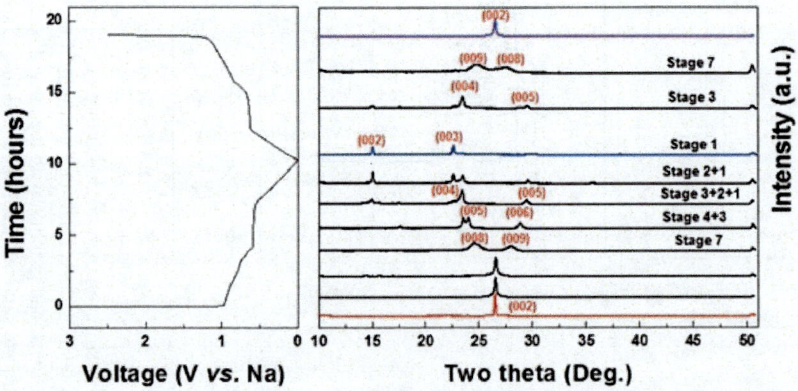

Fig. 2.11 Typical charge/discharge profile and ex-situ XRD patterns using 1 M $NaPF_6$ in DME electrolyte. Reproduced with permission from The Royal Society of Chemistry

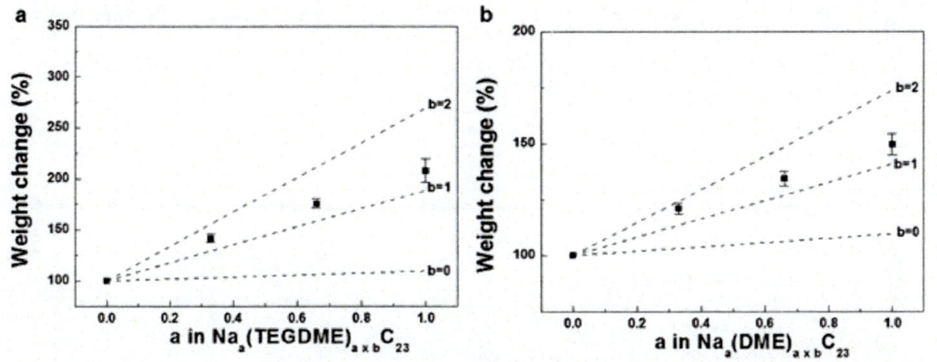

Fig. 2.12 Mass change of the electrodes upon discharge when **a** TEGDME and **b** DME solvents are used. Reproduced with permission from The Royal Society of Chemistry

TEGDME/Na ratio values in GICs are ~1 analogous to the Na-DEGDME electrolyte system (Fig. 2.12). The XRD peak positions of fully discharged samples in Na-TEGDME, Na-DEGDME, and Na-DME systems are almost identical among one another, showing sharp (002) and (003) peaks at 15.2° and 22.9° with c lattice parameters of 11.62 Å (Fig. 2.8b). The observed XRD patterns also match well with simulated XRD patterns from the model structure in Fig. 2.3d. Based on all the results, we deduce that the structures of three GICs are almost identical where the $[Na\text{-}Ether]^+$ complexes are parallel to the graphene layers at one-third and two-third height of the gallery as we calculated. Minor peaks in 11–22° region of the simulated XRD patterns seem to be related to the ordering of $[Na\text{-}DEGDME]^+$ complexes in model structure, as they rise and diminish as the degree of ordering is artificially modified (Fig. 2.13). This supports the presence of the in-plane superstructural ordering of $[Na\text{-}DEGDME]^+$ complexes at highly discharged graphite electrodes shown in the experimental result in Fig. 2.1.

One interesting thing to notice in Fig. 2.8b is that (002)/(003) peak ratios of Na-DME, Na-DEGDME, Na-TEGDME systems are significantly different, despite their discharge capacity and interlayer distance are similar. The intensity of (003) peak was proportional to the length of the solvent (Fig. 2.8c) and it was almost negligible when no intercalants were present. As the peaks in XRD patterns indicates the occupation of electrons in certain space, we made a closer look on the (003) planes in model structure (Fig. 2.3d). In the model GICs, the electron density at (003) planes was determined by the length of the solvent. Therefore, Na-TEGDME which has the largest electron density in threefold repeating motif exhibits the smallest (002)/(003) peak intensity ratio.

Another interesting phenomenon to notice is that the Na storage potential increases with the chain length of solvents. The plateaus were observed at 0.59, 0.65, and 0.77 V (vs. Na) in DME, DEGDME, and TEGDME, respectively (Fig. 2.8d). This solvent dependency is another strong evidence for solvated Na ion intercalation. The solvent-dependent Na storage potential is attributable to the screening effect by the solvent species. Generally, the Na storage potential is described by the energy difference

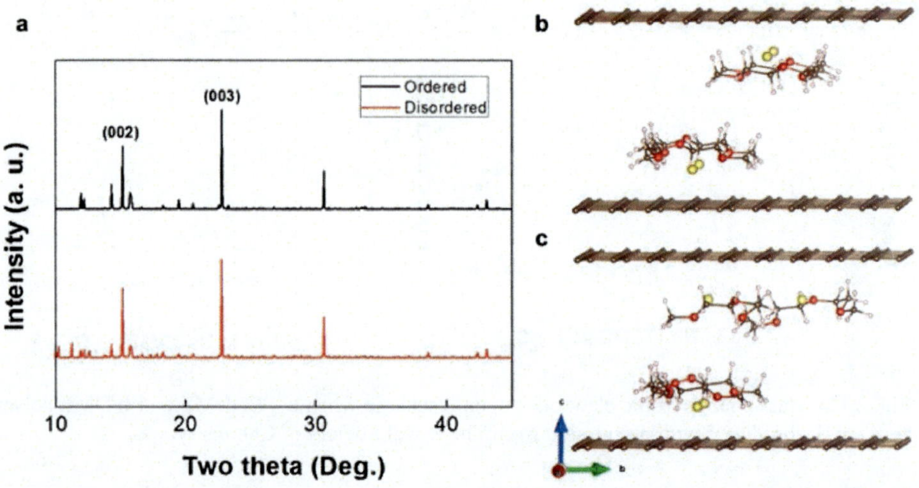

Fig. 2.13 **a** Simulated XRD patterns of two different state of $[Na\text{-}DEGDME]^+$ ordering in graphite host. Structures of ordered, disordered structures used in simulation are described in (**b**), (**c**), respectively. Note that the intensity ratio of (002)/(003) does not change, whereas several minor peaks in 11–22° range emerge or diminish due to the change of $[Na\text{-}DEGDME]^+$ ordering. Reproduced with permission from The Royal Society of Chemistry

between pristine (here, graphite) and intercalated products (here, Na-solvent-graphite compound). In our case, the energetics of intercalated product primarily affect the overall Na storage potential, because the intercalation started from the identical material (graphite). As neutral molecules in the graphite galleries can stabilize the intercalated product by screening the repulsion between positively charged Na ions [31], stronger screening effect would be applied when longer solvent species are introduced, which stabilizes the final discharge product and finally increases Na storage potential (Fig. 2.8e).

2.6 Conclusions

We investigated the solvated ion intercalation of the ternary GIC system of Na-ether-graphite, which has been poorly understood because of its complexity in contrast to the simple binary GIC system of Li-graphite. *Operando* XRD analysis and direct visualization of the formation of the first-stage GICs coupled with DFT calculations provided comprehensive information on solvated Na-ether intercalation into graphite. Despite our comprehensive investigation on the solvated ion intercalation phenomena, it is still unclear why these unique phenomena are observed in using only specific solvents (i.e. DME, DEGDME, TEGDME). While further study is needed, a possible interpretation is that the linear ether solvents strongly solvate Na ions primarily due to the enhanced affinity between solvent and Na ion (so called chelate effect) [32, 33]. The chelate effect describes that the affinity of chelating molecules for a metal ion is enhanced compared to that of a collection of nonchelating (monodentate) molecules for the same metal ion. In this theory, less entropy of disorder is lost when the chelate complex is formed than the complex is formed with monodentate solvents, which results in the enhanced solvation affinity between linear ether solvent molecule and Na ion.

This work also demonstrates that the solvated ion intercalation chemistry can be utilized in energy storage systems based on the high reversibility of the intercalation-induced structural transformations. In energy storage applications, cycle stability that means how many times an electrode is capable of reversibly discharging and re-charging is one of the most important properties. In general, it is known that the volume change affects the cycle performance in the intercalation electrode for rechargeable batteries [34]. Although graphite electrode suffers severe volume change of 347% upon charge/discharge, it exhibits excellent cycle stability (Figs. 2.14 and 2.15). Note that no noticeable structure degradation in both bulk and surface

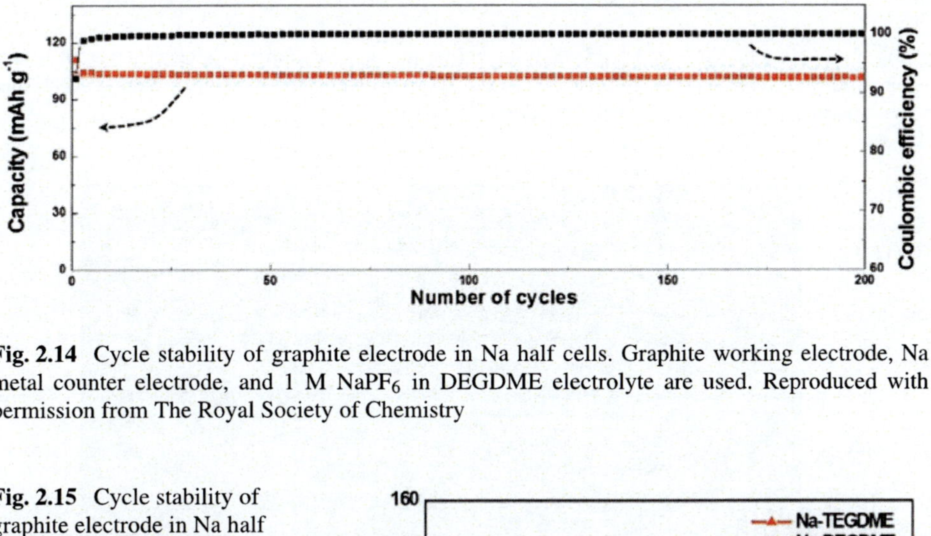

Fig. 2.14 Cycle stability of graphite electrode in Na half cells. Graphite working electrode, Na metal counter electrode, and 1 M $NaPF_6$ in DEGDME electrolyte are used. Reproduced with permission from The Royal Society of Chemistry

Fig. 2.15 Cycle stability of graphite electrode in Na half cells. Graphite working electrode, Na metal counter electrode are used. 1 M $NaPF_6$ in TEGDME, DEGDME, DME electrolytes are used, respectively. Reproduced with permission from The Royal Society of Chemistry

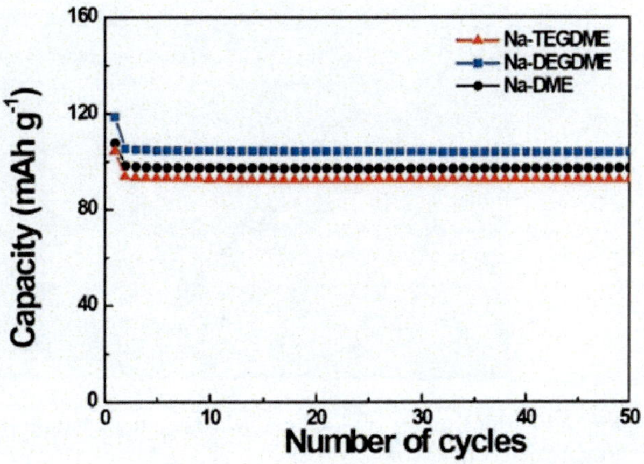

of graphite electrode was observed after 50 cycles in Na-DEGDME electrolyte system (Figs. 2.16 and 2.17). We attribute the excellent reversibility to (i) strong ionic binding between Na and graphene layer preventing from pulverization upon repeated battery cycling (Fig. 2.7) and (ii) the stable solvation complex in the graphite gallery different from propylene carbonate-based electrolytes, wherein intercalated

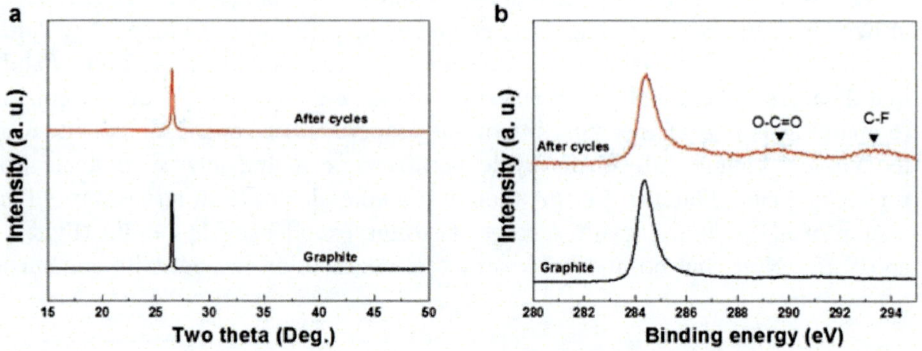

Fig. 2.16 Ex-situ XRD and XPS analyses before and after cycling. **a** XRD patterns of graphite electrode before/after cycling. **b** XPS spectra of graphite electrode before/after cycling. Reproduced with permission from The Royal Society of Chemistry

Fig. 2.17 SEM image of graphite electrode after cycling. Reproduced with permission from The Royal Society of Chemistry

propylene carbonate solvents are decomposed to gas phases and exfoliate graphene layers [35–37].

There have also been several reports on the electrochemical storage of cations in host materials which can be enabled by the aid of proper solvents similar with our system, indicative of the growing importance of the role of electrolyte solvents [16, 38, 39]. In this respect, we believe our work helps expand the pool of electrode materials from the binary systems of guest ion-host materials to versatile ternary systems of guest ion-solvent-host materials. These findings will lead us to further investigation towards the energy storage materials using solvated ion intercalation.

References

1. L.B. Ebert, Intercalation compounds of graphite. Annu. Rev. Mater. Sci. **6**, 181–211 (1976)
2. M.S. Dresselhaus, G. Dresselhaus, Intercalation compounds of graphite. Adv. Phys. **30**, 139–326 (1981)
3. T.E. Weller, M. Ellerby, S.S. Saxena, R.P. Smith, N.T. Skipper, Superconductivity in the intercalated graphite compounds C_6Yb and C_6Ca. Nat. Phys. **1**, 39–41 (2005)
4. K. Sugihara, *c*-axis conduction in graphite intercalation compounds. Phys. Rev. B **37**, 4752–4759 (1988)
5. K. Sugihara, c-axis conductivity and thermoelectric power in graphite intercalation compounds. Phys. Rev. B **29**, 5872–5877 (1984)
6. H. Cheng, X. Sha, L. Chen, A.C. Cooper, M.-L. Foo, G.C. Lau, W.H. Bailey III, G.P. Pez, An enhanced hydrogen adsorption enthalpy for fluoride intercalated graphite compounds. J. Am. Chem. Soc. **131**, 17732–17733 (2009)
7. A. Grüneis, C. Attaccalite, A. Rubio, D.V. Vyalikh, S.L. Molodtsov, J. Fink, R. Follath, W. Eberhardt, B. Büchner, T. Pichler, Angle-resolved photoemission study of the graphite intercalation compound KC_8: a key to graphene. Phys. Rev. B **80**, 075431 (2009)
8. K. Kanetani, K. Sugawara, T. Sato, R. Shimizu, K. Iwaya, T. Hitosugi, T. Takahashi, Ca intercalated bilayer graphene as a thinnest limit of superconducting C_6Ca. Proc. Natl. Acad. Sci. **109**, 19610–19613 (2012)
9. M. Conti-Ramsden, K. Nkrumah-Amoako, N. Brown, E.L. Roberts, The oxidation of aqueous thiols on a graphite intercalation compound adsorbent. Adsorption **19**, 989–996 (2013)
10. S. Park, R.S. Ruoff, Chemical methods for the production of graphenes. Nat. Nanotechnol. **4**, 217–224 (2009)
11. Y. Nishi, Lithium ion secondary batteries; past 10 years and the future. J. Power Sources **100**, 101–106 (2001)
12. D. Aurbach, Y. Ein-Eli, The study of Li-graphite intercalation processes in several electrolyte systems using in situ X-ray diffraction. J. Electrochem. Soc. **142**, 1746–1752 (1995)
13. K. Nobuhara, H. Nakayama, M. Nose, S. Nakanishi, H. Iba, First-principles study of alkali metal-graphite intercalation compounds. J. Power Sources **243**, 585–587 (2013)
14. Z. Wang, S.M. Selbach, T. Grande, Van der Waals density functional study of the energetics of alkali metal intercalation in graphite. RSC Adv. **4**, 4069–4079 (2014)
15. B. Jache, P. Adelhelm, Use of graphite as a highly reversible electrode with superior cycle life for sodium-ion batteries by making use of co-intercalation phenomena. Angew. Chem. Int. Ed. **53**, 10169–10173 (2014)
16. H. Kim, J. Hong, Y.-U. Park, J. Kim, I. Hwang, K. Kang, Sodium storage behavior in natural graphite using ether-based electrolyte systems. Adv. Funct. Mater. **25**, 534–541 (2015)
17. V.A. Sethuraman, L.J. Hardwick, V. Srinivasan, R. Kostecki, Surface structural disordering in graphite upon lithium intercalation/deintercalation. J. Power Sources **195**, 3655–3660 (2010)

18. T. Zheng, J.R. Dahn, The effect of turbostratic disorder on the staging transitions in lithium intercalated graphite. Synth. Met. **73**, 1–7 (1995)
19. A. Funabiki, M. Inaba, T. Abe, Z. Ogumi, Stage transformation of lithium-graphite intercalation compounds caused by electrochemical lithium intercalation. J. Electrochem. Soc. **146**, 2443–2448 (1999)
20. K. Persson, Y. Hinuma, Y.S. Meng, A. Van der Ven, G. Ceder, Thermodynamic and kinetic properties of the Li-graphite system from first-principles calculations. Phys. Rev. B **82**, 125416 (2010)
21. M.J. Frisch, G.W. Trucks, H.B. Schlegel, G.E. Scuseria, M.A. Robb, J.R. Cheeseman, G. Scalmani, V. Barone, B. Mennucci, G.A. Petersson, H. Nakatsuji, M. Caricato, X. Li, H.P. Hratchian, A.F. Izmaylov, J. Bloino, G. Zheng, J.L. Sonnenberg, M. Hada, M. Ehara, K. Toyota, R. Fukuda, J. Hasegawa, M. Ishida, T. Nakajima, Y. Honda, O. Kitao, H. Nakai, T. Vreven, J.A. Montgomery Jr., J.E. Peralta, F. Ogliaro, M.J. Bearpark, J. Heyd, E.N. Brothers, K.N. Kudin, V.N. Staroverov, R. Kobayashi, J. Normand, K. Raghavachari, A.P. Rendell, J.C. Burant, S.S. Iyengar, J. Tomasi, M. Cossi, N. Rega, N.J. Millam, M. Klene, J.E. Knox, J.B. Cross, V. Bakken, C. Adamo, J. Jaramillo, R. Gomperts, R.E. Stratmann, O. Yazyev, A.J. Austin, R. Cammi, C. Pomelli, J.W. Ochterski, R.L. Martin, K. Morokuma, V.G. Zakrzewski, G.A. Voth, P. Salvador, J.J. Dannenberg, S. Dapprich, A.D. Daniels, Ö. Farkas, J.B. Foresman, J.V. Ortiz, J. Cioslowski, D.J. Fox, *Gaussian 09* (Gaussian, Inc., Wallingford, CT, 2009)
22. G. Kresse, J. Furthmüller, Efficiency of ab-initio total energy calculations for metals and semiconductors using a plane-wave basis set. Comput. Mater. Sci. **6**, 15–50 (1996)
23. P.E. Blöchl, Projector augmented-wave method. Phys. Rev. B **50**, 17953–17979 (1994)
24. J.P. Perdew, K. Burke, M. Ernzerhof, Generalized gradient approximation made simple. Phys. Rev. Lett. **77**, 3865–3868 (1996)
25. S. Grimme, Semiempirical GGA-type density functional constructed with a long-range dispersion correction. J. Comput. Chem. **27**, 1787–1799 (2006)
26. J.A. Seel, J.R. Dahn, Electrochemical intercalation of PF_6 into graphite. J. Electrochem. Soc. **147**, 892–898 (2000)
27. W. Sirisaksoontorn, A.A. Adenuga, V.T. Remcho, M.M. Lerner, Preparation and characterization of a tetrabutylammonium graphite intercalation compound. J. Am. Chem. Soc. **133**, 12436–12438 (2011)
28. Y. Yamada, K. Usui, C.H. Chiang, K. Kikuchi, K. Furukawa, A. Yamada, General observation of lithium intercalation into graphite in ethylene-carbonate-free superconcentrated electrolytes. ACS Appl. Mater. Interfaces **6**, 10892–10899 (2014)
29. H. Moon, R. Tatara, T. Mandai, K. Ueno, K. Yoshida, N. Tachikawa, T. Yasuda, K. Dokko, M. Watanabe, Mechanism of Li ion desolvation at the interface of graphite electrode and glyme–Li salt solvate ionic liquids. J. Phys. Chem. C **118**, 20246–20256 (2014)
30. G. Yoon, D.-H. Seo, K. Ku, J. Kim, S. Jeon, K. Kang, Factors affecting the exfoliation of graphite intercalation compounds for graphene synthesis. Chem. Mater. **27**, 2067–2073 (2015)
31. N.I. Kovtyukhova, Y. Wang, A. Berkdemir, R. Cruz-Silva, M. Terrones, V.H. Crespi, T.E. Mallouk, Non-oxidative intercalation and exfoliation of graphite by Brønsted acids. Nat. Chem. **6**, 957–963 (2014)
32. G. Schwarzenbach, Der Chelateffekt. Helv. Chim. Acta **35**, 2344–2359 (1952)
33. R.D. Hancock, A.E. Martell, The chelate, cryptate and macrocyclic effects. Comments Inorg. Chem. **6**, 237–284 (1988)
34. Y.-U. Park, D.-H. Seo, H.-S. Kwon, B. Kim, J. Kim, H. Kim, I. Kim, H.-I. Yoo, K. Kang, A new high-energy cathode for a Na-ion battery with ultrahigh stability. J. Am. Chem. Soc. **135**, 13870–13878 (2013)
35. H. Buqa, A. Würsig, D. Goers, L.J. Hardwick, M. Holzapfel, P. Novák, F. Krumeich, M.E. Spahr, Behaviour of highly crystalline graphites in lithium-ion cells with propylene carbonate containing electrolytes. J. Power Sources **146**, 134–141 (2005)
36. M.R. Wagner, J.H. Albering, K.C. Moeller, J.O. Besenhard, M. Winter, XRD evidence for the electrochemical formation of $Li^+ (PC)_y C_n^-$ in PC-based electrolytes. Electrochem. Commun. **7**, 947–952 (2005)

37. A.N. Dey, B.P. Sullivan, The electrochemical decomposition of propylene carbonate on graphite. J. Electrochem. Soc. **117**, 222–224 (1970)
38. H. Pan, X. Lu, X. Yu, Y.-S. Hu, H. Li, X.-Q. Yang, L. Chen, Sodium storage and transport properties in layered $Na_2Ti_3O_7$ for room-temperature sodium-ion batteries. Adv. Energy Mater. **3**, 1186–1194 (2013)
39. Z. Hu, Z. Zhu, F. Cheng, K. Zhang, J. Wang, C. Chen, J. Chen, Pyrite FeS_2 for high-rate and long-life rechargeable sodium batteries. Energy Environ. Sci. **8**, 1309–1316 (2015)

Chapter 3
Conditions for Reversible Na Intercalation in Graphite

3.1 Introduction

The soaring energy demand from the rapidly growing industries is impossible to be met by the consumption of the fossil fuels, which are not only harmful to our environment but also have limited supply. Therefore, the utilization of eco-friendly energy sources and their storage systems have recently drawn a great attention. Na-ion batteries (NIBs) system is an attractive candidate for large-scale energy storage system because the ready availability of Na could make the system cost-effective comparing to the current Li-ion batteries (LIBs) [1–4]. However, choosing an appropriate anode material has been one of the greatest challenges in NIBs because early experiments have observed that graphite, which is widely utilized in LIBs for its low operation voltage, decent reversible capacity and low cost, could not work as a Na intercalation host [5, 6].

Theoretical studies have also confirmed the instability of Na intercalation into graphite [7–10]. Nobuhara et al. found that the stage-1 Na graphite intercalation compounds (GICs) such as NaC_6 and NaC_8 are unstable by the GGA-PBE calculation and attributed the stress caused by the stretch of C–C bond lengths to the reason for unstable Na-GICs [7]. Okamoto calculated that the formation of NaC_6 and NaC_8 is thermodynamically unfavorable regardless of the exchange-correlation functionals, and claimed that the high redox potential of Na/Na^+ is accountable for this behaviour [8]. Recently, Wang et al. [9] and Liu et al. [10] conducted more detailed investigation to understand the underlying reason for the Na instability in the graphite host by the de-convolution of the formation energy according to the Hess' law. Formation energy of GICs is categorized as following three factors: formation energies upon reconstruction of (i) alkali metal and (ii) graphite framework, and (iii) the remaining

The content of this chapter has been published in *Advanced Energy Materials.* [Yoon, G. et al., *Advanced Energy Materials* **2017**, 7(2), 1601519.] Reproduced by permission of 2016 WILEY-VCH Verlag GmbH & Co. KGaA, Weinheim.

G. Yoon, *Theoretical Study on Graphite and Lithium Metal as Anode Materials for Next-Generation Rechargeable Batteries*, Springer Theses,
https://doi.org/10.1007/978-981-13-8914-6_3

formation energy. Although they pointed out that the factor (iii) is responsible for the Na instability in graphite host, the underlying reason is not clear yet because the factor (iii) in these works is simply the sum of all energy contributions other than the reconstruction of alkali metal and graphite by definition.

The thermodynamic instability of Na-GICs formation has led researchers to the development of an alternative way for graphite utilization. Recently, the formation of ternary GICs by Na-solvent co-intercalation has been reported, which overcomes the unstable formation of binary Na-GICs and enables the utilization of graphite as an anode for NIBs [11–15]. Nevertheless, it is still poorly understood why only specific solvent species, such as linear ethers and their derivatives, make the reversible Na-solvent co-intercalation phenomenon possible. Unveiling the underlying reason for the Na-solvent co-intercalation behavior would broaden our understanding on co-intercalation chemistry in graphite host and provide insights for the design of better system to utilize graphite as a host material for rechargeable batteries.

In this work, we consider the possible energy contributions of AM-GICs formation (AM = Li, Na, K, Rb, Cs) without following Hess' law, in order to ensure that the pure energy contributions are extracted without being mixed with other factors. We found that the weak interaction between Na ion and graphene layer is the underlying reason for the instability of binary Na-GICs. We further demonstrate that the Na storage in graphite is possible through co-intercalation phenomenon because the neutral solvent molecules can minimize the direct interaction between Na ions and graphene layers. In addition, we investigate the solvent dependency on Na-solvent co-intercalation behavior and suggest determining factors for reversible Na-solvent co-intercalation. We believe this work will provide new insights in designing novel intercalation based energy storage system including graphite host materials.

3.2 Computational Details

All information on geometries, formation energies and electronic structures of alkali metal-intercalated graphite intercalation compounds (AM-GICs) were obtained by performing density functional theory (DFT) calculation using Vienna Ab initio Simulation Package (VASP) [16]. Exchange-correlation energies were described using spin-polarized generalized gradient approximation (GGA) parametrized by Perdew–Burke–Ernzerhof (PBE) [17]. We used projector augmented wave (PAW) pseudopotentials as implemented in VASP [18, 19]. A plane-wave basis set was used with an energy cutoff of 500 eV and geometric relaxations were performed until the remaining force in the system converges within 0.02 eV $Å^{-1}$. To describe the van der Waals force between graphene layers in AM-GICs, we adopted the method suggested by Grimme et al. (DFT-D3) [20]. The validity of DFT-D3 scheme is confirmed by the calculation of graphite, LiC_6, KC_8 and comparing their structural properties to experimental values (Table 3.1). Structures, formation energies and HOMO/LUMO levels of $[\text{Na-solvent}_x]^+$ complexes were calculated using Gaussian 09 code [21]. All molecules were optimized with B3LYP/6-311G (3df) level [22, 23].

Table 3.1 Comparison of c-lattice parameters obtained by experiments and PBE-D3 calculation

	$d_{graphite}$ (Å)	d_{LiC_6} (Å)	d_{KC_8} (Å)
DFT-D3	3.46	3.76	5.30
Exp.	3.36	3.71 [24]	5.35 [25]

Reproduced with permission from WILEY-VCH

3.3 Unstable Na Intercalation in Graphite

To investigate the thermodynamic stability of stage 1 AM-GICs, the formation energies of MC_6 and MC_8 (M = Li, Na, K, Rb, Cs) are calculated as depicted in Fig. 3.1a. Formation energy E_f is defined as

$$E_f = E_{MC_x} - x \times E_g - E_M$$

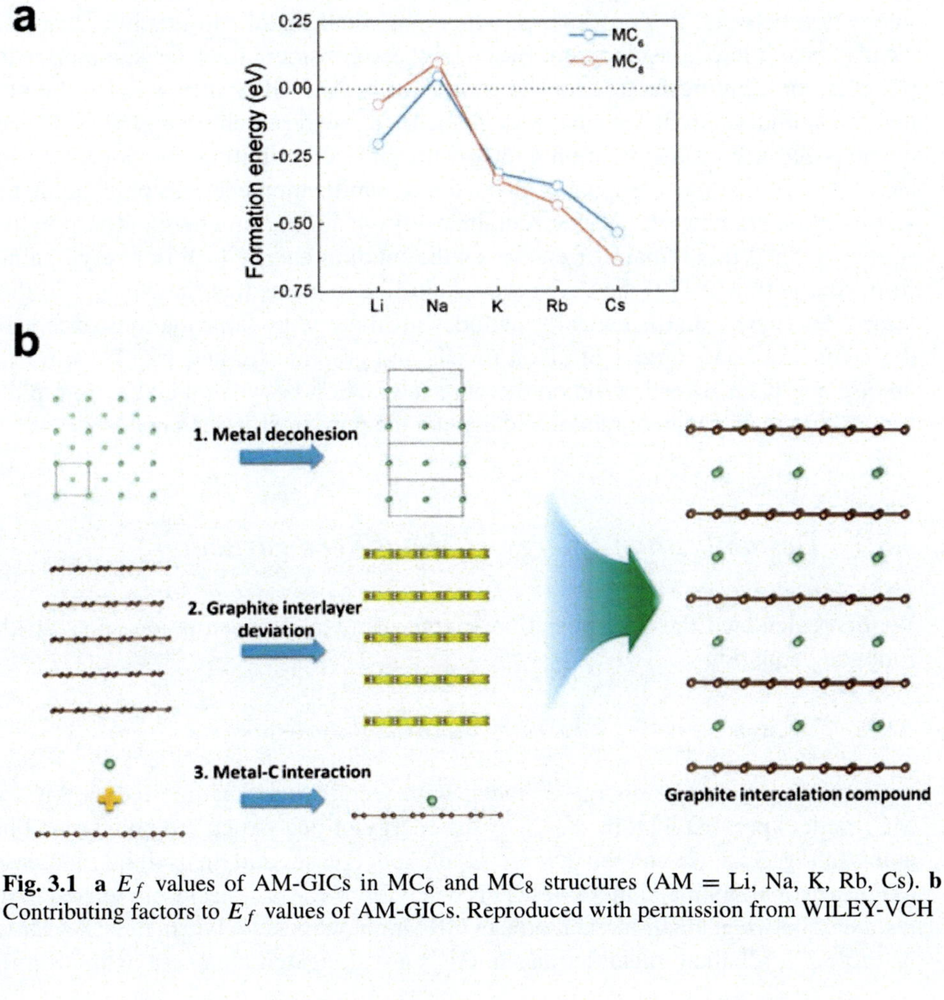

Fig. 3.1 **a** E_f values of AM-GICs in MC_6 and MC_8 structures (AM = Li, Na, K, Rb, Cs). **b** Contributing factors to E_f values of AM-GICs. Reproduced with permission from WILEY-VCH

where E_{MC_x} denotes the energy of MC_x formula unit, E_g is the energy of graphite per carbon atom and E_M is the energy of solid alkali metal per metal atom. The initial structures (i.e. the interlayer ordering of graphene layers and metal layers) of MC_6 and MC_8 are adopted from the experimental reports on LiC_6 [24] and KC_8 [26]. As experimentally reported, the formation of binary Na-GICs is thermodynamically unfavorable, whereas other alkali metals form stable stage 1 binary GICs [25]. We also note that the composition of stage 1 AM-GICs differs among alkali metals. As intercalant becomes larger, MC_8 composition is preferred; Li prefers to form LiC_6, whereas K, Rb and Cs prefer to form KC_8, RbC_8 and CsC_8, respectively. The detailed reason for this trend will be discussed in the following sections.

For a better understanding on the Na instability in the formation of AM-GICs, we categorize the possible candidates which are expected to have major effects on the formation energy of AM-GICs. Since the AM-GICs are composed of the alternating metal layers and graphene layers, it is natural to expect that the structural change of alkali metals and graphite from their ground states greatly contributes to the total formation energies. In addition, the formation of AM-GIC involves the interaction between its constituents, which are alkali metal and graphite, therefore we also take it into consideration. Based on these considerations, we summarize the possible contributing factors for the formation of AM-GICs in Fig. 3.1b. We note that the similar approaches have been made by a few research groups [9, 10]. They decomposed the total formation energy into three contributing components using Hess' law where three components were the destabilization energies of alkali metals and graphite framework and the remainder of the formation energy. The previous studies defined this remainder energy as the binding energy [10] or energy gained from charge transfer [9], however, we would like to stress that the concept of their remainder energy mathematically includes all the energy contributions except for the reconstruction energies of alkali metals and graphite framework. Thus, for the precise and clear investigation on the pure interaction between metal and graphite, we constructed a separate model to calculate the pure interaction energy.

3.3.1 Destabilization Energy of Metal Reconstruction

We first calculated the destabilization energy of metal reconstruction ($E_{d,metal}$) by following equation:

$$E_{d,metal} = E_{m,GIC} - E_{m,bcc}$$

Here, $E_{m,GIC}$ is the energy of metal with reconstructed lattice (either MC_6 or MC_8 lattice) per metal atom, $E_{m,bcc}$ is the energy of bcc-structured alkali metal per atom. In Fig. 3.2a, we can see that $E_{d,metal}$ first decreases and then slightly increases as the size of alkali metal increases. This behavior of $E_{d,metal}$ is closely related to the distances between alkali metal atoms in GIC frameworks and bcc metals. As shown in Table 3.2, all alkali metal lattices in GICs are elongated along c-axis in a similar

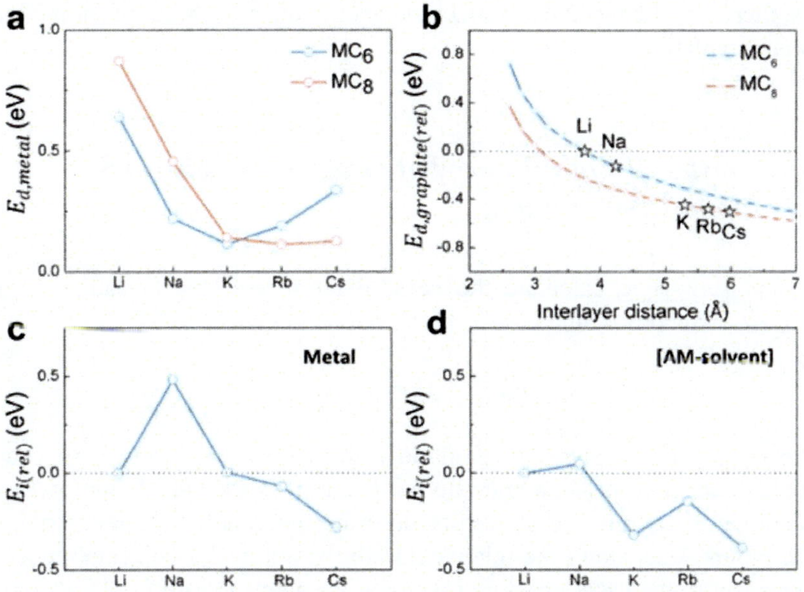

Fig. 3.2 **a** $E_{d,metal}$ values upon lattice reconstruction from bcc structure to MC_6 and MC_8 structures. **b** Relative $E_{d,graphite}$ upon AM intercalation and charge transfer. **c** Relative E_i between AM and a single layer of graphene. **d** Relative E_i between AM-solvent complexes and a single layer of graphene. The energy scales in (**b**), (**c**), and (**d**) are referenced to the values obtained for Li. Reproduced with permission from WILEY-VCH

Table 3.2 Bond lengths of AMs in bcc lattice d_{bcc} and comparison of d_{bcc} to the AM lattices in MC_6 and MC_8 structures in both *ab*- and *c*-directions

Metal	d_{bcc} (Å)	$d_{MC_6,c}$ (Å)	$d_{MC_8,c}$ (Å)	$d_{MC_6,ab}$ (Å)	$d_{MC_8,ab}$ (Å)
Li	2.97	3.76	3.76		
Na	3.63	4.24	4.5		
K	4.57	5.25	5.3	4.32	4.93
Rb	4.90	5.6	5.66		
Cs	5.32	5.97	5.98		

Reproduced with permission from WILEY-VCH

degree because of the expansion of graphite host upon alkali metal intercalation. However, alkali metals in AM-GICs have the same distance along *ab*-plane (4.32 Å and 4.93 Å in MC_6 and MC_8 structure, respectively) because of the fixed intercalation sites. Therefore, the degree of metal lattice mismatch along *ab*-plane determines the $E_{d,metal}$, and K, Rb shows the smallest $E_{d,metal}$ in MC_6, MC_8 structure, respectively. This result explains the trend where larger intercalants prefer to form MC_8 (Fig. 3.1a). Since the MC_8 structure has large interatomic distances between metal sites, larger metals such as Rb, Cs prefer to form MC_8 and vice versa. However, the trend of $E_{d,metal}$ exhibits no anomaly for the case of Na intercalation in either structure of

GICs, therefore the metal destabilization energy does not account for the instability of Na intercalation.

3.3.2 Destabilization Energy of Graphite Framework upon Intercalation

Our next step is to calculate the destabilization energy of graphite framework $E_{d,graphite}$, which is defined as

$$E_{d,graphite} = E_{g,GIC} - E_g$$

where $E_{g,GIC}$ is the energy of graphite with reconstructed lattice per C atom and E_g is the energy of pristine graphite per C atom. In the calculation of $E_{g,GIC}$, we injected electrons into the system to describe the charged graphite framework in GICs. Figure 3.2b shows the relative destabilization energy of graphite framework with respect to the LiC_6 structure assuming the electrons are fully transferred from alkali metal to graphite regardless of the metal species. We can see that the graphite destabilization energy decreases as the interlayer distance between graphene layers increases according to the fundamental electrostatics law that the electrostatic repulsion between charged species is inverse proportional to the distance between them. This monotonous trend of $E_{d,graphite}$ could also not be a reason for peculiar Na instability.

3.3.3 Local Interaction Between Alkali Metal Ions and the Graphite Framework

The pure interaction between alkali metal and graphite framework is also calculated with a simplified model including a single layer of graphite and a metal atom (Fig. 3.3). The interaction energy E_i is defined as

$$E_i = E_{m+graphene} - E_m - E_{graphene}$$

where $E_{m+graphene}$ is the energy of the metal adsorbed graphene, E_m is the energy of metal atom and $E_{graphene}$ is the energy of a single layer graphene per C atom. Relative values of E_i with respect to $E_{i,Li}$ are plotted in Fig. 3.2c. Unlike the previously investigated components ($E_{d,metal}$ and $E_{d,graphite}$) which show the monotonous trend towards the stable formation of AM-GICs, Na ion binding to graphene layer is peculiarly unstable by ~0.5 eV comparing to other alkali metals [27]. This observation well corresponds to the trend of the formation energy E_f of GICs (Fig. 3.1a), thus, the interaction between metal ions and graphene layers is expected to determine the

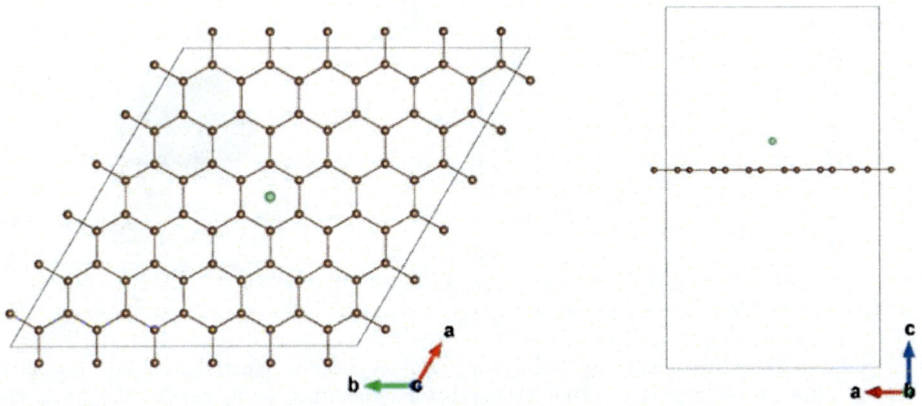

Fig. 3.3 Model structure for the calculation of pure interaction between AM and the graphene layer. A vacuum slab of ~15 Å was used. Reproduced with permission from WILEY-VCH

E_f of GICs and the instability of Na-GIC formation. To understand this abnormal phenomenon, we observed the behavior of charges by performing Bader charge analysis and compared the amount of charge transferred for all alkali metal species (Table 3.3). All alkali metals exhibit Bader charges of more than +0.89, indicating that the interaction between a metal atom and a graphene layer is purely ionic. Despite the high $E_{i,Na}$, the amount of charge transfer falls in a range of 0.89–0.91 electrons per atom regardless of the metal species, which could not account for the high $E_{i,Na}$. We further investigated the distribution of the transferred charges by calculating the plane-averaged charge densities (Fig. 3.4). However, the distribution of additional charges from alkali metals is also similar to each other. We note that the trends of atomization energy and ionization energy are also monotonous for alkali metal series. Our calculations on the energy components hitherto clearly indicate that the peculiarly weak Na-graphene interaction is the underlying reason for the unstable Na-GIC formation, while the Na anomaly of E_i could not be clearly explained, and speculated to be related to the intrinsic property of Na.

Table 3.3 Bader charges of AMs in metal-adsorbed graphene

	Li	Na	K	Rb	Cs
Bader charge	+0.90	+0.89	+0.91	+0.91	+0.91

Reproduced with permission from WILEY-VCH

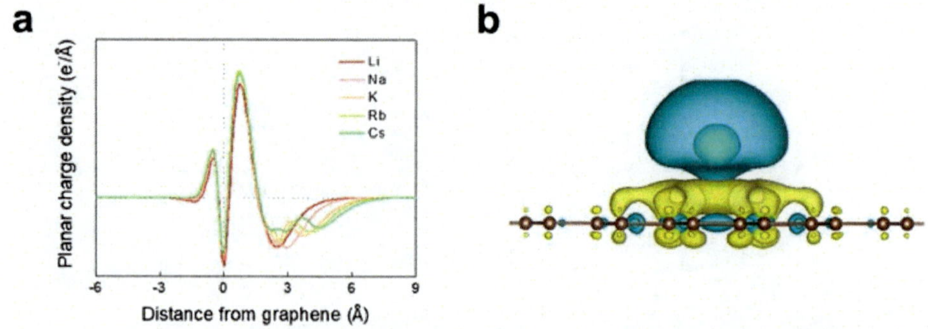

Fig. 3.4 **a** Plane-averaged charge density of metal-adsorbed graphene. Because of the nature of AMs, different characteristics of the charge distributions, such as the peak of the planar charge density and distance from graphene at maximum charge density, are observed. **b** Charge distribution of Li-adsorbed graphene. All the AM-adsorbed graphene samples exhibited identical charge distributions. Isosurface level is set to 0.005 $e^{-}/Å^{3}$. Reproduced with permission from WILEY-VCH

3.3.4 Mitigating the Unfavorable Local Interaction Between Na and Graphene Layers

Based on our calculations, we can expect that graphite can accommodate Na ions in its framework if the interaction between Na ions and graphene layers is minimized. One of the plausible approaches is the screening of Na ion from graphite framework to prevent the direct Na-graphite interaction. The interaction between alkali metal and graphite would be diluted by the screening moieties and the peculiarly unstable Na-graphite interaction could be relieved. In this respect, recent reports on the reversible Na intercalation behavior in graphite using the Na-solvent co-intercalation phenomenon [11–15] can be also explained by the screening effect of the intercalated solvent molecules which prevents the direct Na-graphite interaction. In order to elucidate the effect of Na-solvent co-intercalation, we calculated the interaction energy E_i of AM-solvent complex and graphene layer (Fig. 3.2d). Diethylene glycol dimethyl ether (DEGDME) is used as solvent because it was experimentally reported to be reversibly co-intercalated in graphite with Na. Unlike the trend of E_i between pure alkali metal and graphene layer (Fig. 3.2c), Na-solvent complex does not exhibit noticeable instability in an interaction with graphene layer. Bulky solvent molecule effectively screens the alkali metal ions and the peculiar behavior of Na is diluted as expected. This observation is further supported by the calculation of plane-averaged charge density where the charge distribution is almost identical for all AM-solvent complexes (Fig. 3.5). Due to the screening effect of solvent molecule, individual characteristics of charge densities of alkali metal ions described in Fig. 3.4 are mostly lost, resulting in the almost identical charge distribution and finally leading to the Na-solvent co-intercalation in graphite.

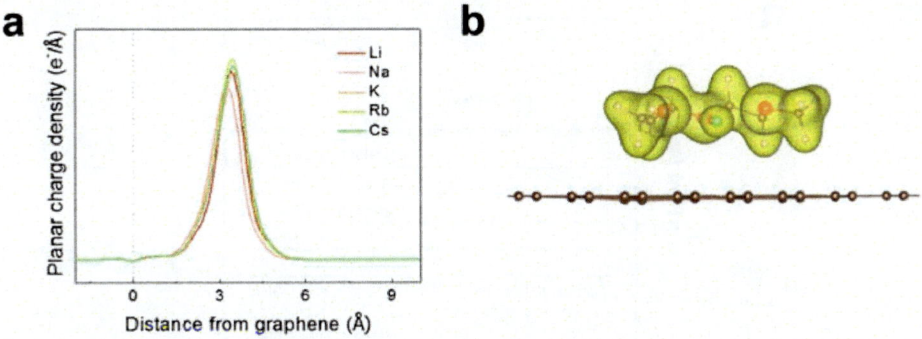

Fig. 3.5 **a** Plane-averaged charge density of [AM-DEGDME]-adsorbed graphene. Because of the screening effect of the solvent molecules, similar charge distributions were observed for all the AMs. **b** Charge distribution of [Li-DEGDME]-adsorbed graphene. All the [AM-DEGDME]-adsorbed graphene samples exhibited identical charge distributions. Isosurface level is set to 0.03 $e^{-}/Å^{3}$. Reproduced with permission from WILEY-VCH

3.4 Conditions of Solvents for Reversible Na Intercalation into Graphite

3.4.1 Solvent Dependency on Reversible Na-Solvent Co-intercalation Behavior

In previous sections, we successfully demonstrated that the inclusion of the screening moieties such as solvent molecules could negate the peculiar interaction between Na and graphite. However, this approach does not always guarantee the reversible intercalation into graphite. The experimental reports on reversible intercalation of Na-solvent complexes into graphite has been limited to a certain range of solvent species such as linear ethers and their derivatives [11–15]. Our preliminary experiment on the Na-solvent co-intercalation behavior using a various category of solvent species (linear/cyclic ethers and carbonates; see Fig. 3.6b for details) also suggests that the co-intercalation into graphite is only practicable with linear ethers as described in Fig. 3.6a. While reduction activity was clearly observed with propylene carbonate (PC) electrolyte, no oxidation occurred, indicative of irreversible intercalation and exfoliation of graphite as described in Scanning Electron Microscopy (SEM) image (Fig. 3.7), which is similar to Li-PC electrolyte system [28]. This solvent dependency on co-intercalation behavior is also reported in Li-graphite system, where linear ethers reversibly intercalate into graphite with Li, while carbonate-based solvents either result in the pure Li intercalation or the graphite exfoliation [29–31]. Previous studies so far suggested that the interaction between Li and solvent, i.e., Li solvation energy or donor number of solvents determine the co-intercalation behavior because the high stability of solvated Li prevents the desolvation after the Li-solvent co-intercalation [30, 32, 33]. However, direct comparison between the donor number of solvents (Table 3.4) and the electrochemical activity does not give a clear tendency in

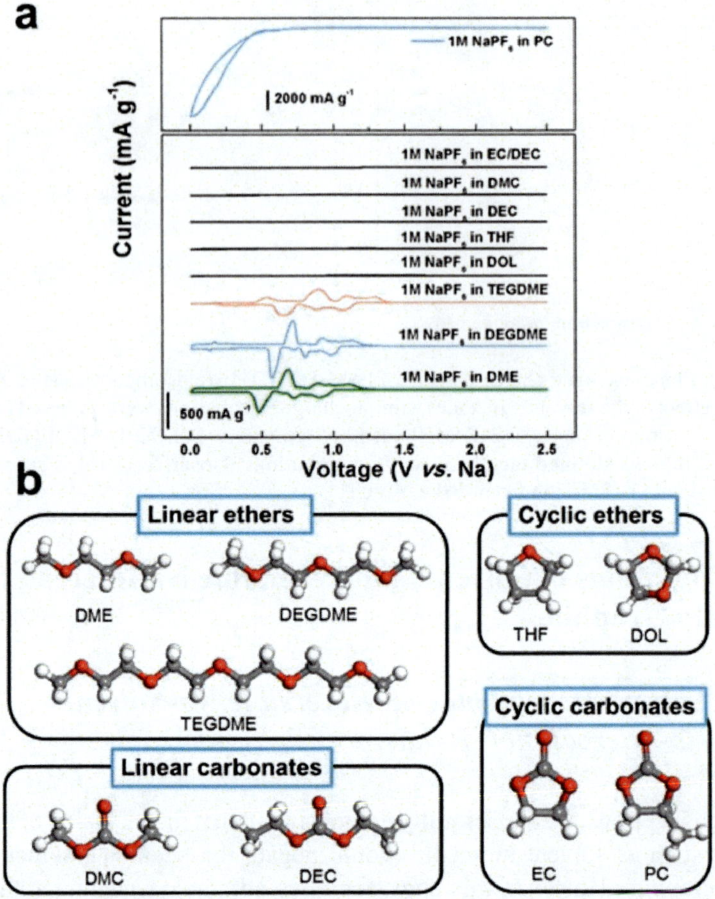

Fig. 3.6 **a** Cyclic voltammetry curves of various Na-solvent systems; 1 M $NaPF_6$ salts were used for all tests. **b** The solvent species used to investigate the solvent dependency of the co-intercalation phenomenon (DME: ethylene glycol dimethyl ether, DEGDME: diethylene glycol dimethyl ether, TEGDME: tetraethylene glycol dimethyl ether, THF: tetrahydrofuran, DOL: dioxolane, DMC: dimethyl carbonate, DEC: diethyl carbonate, EC: ethylene carbonate, PC: propylene carbonate). Reproduced with permission from WILEY-VCH

Na-graphite system. Cyclic ethers have higher donor number than linear ethers, but they exhibit no co-intercalation phenomenon (Fig. 3.6a). Therefore, in order to better understand the underlying reason of this solvent dependency on co-intercalation behavior in Na-graphite system, it is necessary to investigate the stabilities of Na-solvent complexes with a various category of solvent species.

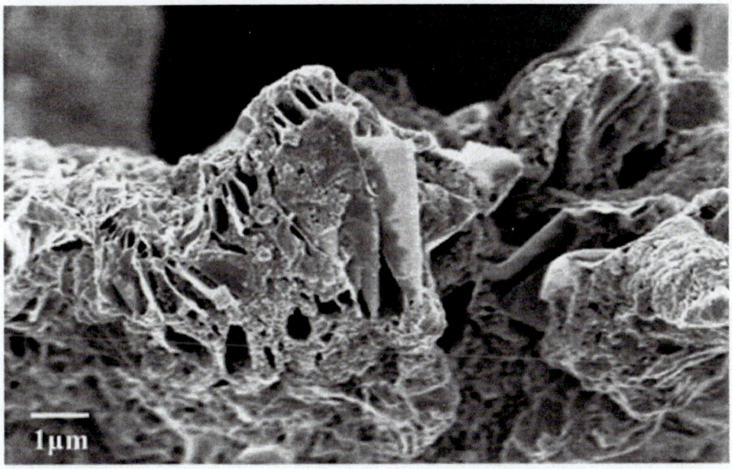

Fig. 3.7 SEM image of graphite after discharge in 1 M $NaPF_6$ in PC solvent. The cleavages observed in the sample indicate the exfoliation of graphite. Reproduced with permission from WILEY-VCH

Table 3.4 Material properties of investigated solvents

Type	Linear carbonate		Cyclic carbonate		Linear ether			Cyclic ether	
Solvent	DMC	DEC	EC	PC	DME	DEG DME	TEG DME	THF	DOL
# of oxygen	3	3	3	3	2	3	5	1	2
Dielectric constant	3.1	2.8	90.5 (40 °C)	64.9	7.0	7.4	7.7	7.6	7.1
Donor number	17.2	16	16.4	15.1	24	19.2	16.6	20	21.2

Reproduced with permission from WILEY-VCH

3.4.2 Thermodynamic Stability of Na-Solvent Complex

We first calculated the solvation energy E_s to compare the thermodynamic stability of Na-solvent complexes in various solvents. E_s was defined as

$$E_s = E_{[Na-solvent_x]^+} - x \times E_{solvent} - E_{Na^+}$$

where $E_{[Na-solvent_x]^+}$ is the energy of Na-solvent complex, $E_{solvent}$ is the energy of solvent molecule and E_{Na^+} is the energy of Na ion. Since the solvation number x could vary in dynamic situation in electrolyte, various solvation number x was applied to each solvent species and the maximum solvation number was adopted from experimental observations in Li solvation chemistry [34–36]. We note that the interaction with anion in Na salts is neglected because the experimental researches on ternary Na-solvent GICs reported that the intercalant is composed of only Na

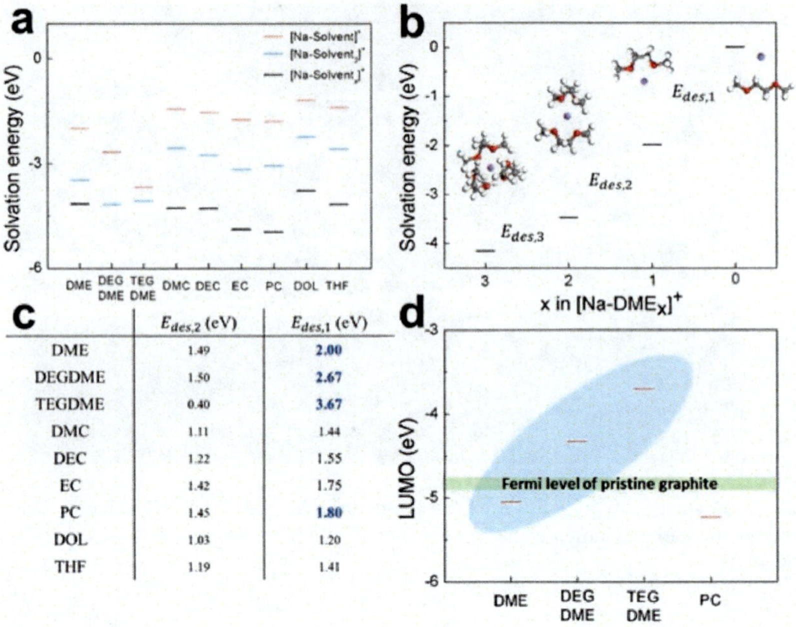

	$E_{des,2}$ (eV)	$E_{des,1}$ (eV)
DME	1.49	**2.00**
DEGDME	1.50	**2.67**
TEGDME	0.40	**3.67**
DMC	1.11	1.44
DEC	1.22	1.55
EC	1.42	1.75
PC	1.45	**1.80**
DOL	1.03	1.20
THF	1.19	1.41

Fig. 3.8 **a** E_s of Na-solvent complexes. **b** Energy diagram of $[\text{Na-DME}_x]^+$ desolvation process. **c** E_{des} values of Na-solvent complexes. The E_{des} values of potential candidates for feasible co-intercalation are highlighted in bold. **d** LUMO levels of $[\text{Na-solvent}]^+$ complexes. The Fermi level of graphite is shaded green. Reproduced with permission from WILEY-VCH

and solvent, and the intercalation behavior was independent on the anion species [12]. Figure 3.8a shows the calculated E_s of all investigated Na-solvent complexes. When Na ion is solvated by single solvent molecule, stronger solvation is observed between Na and solvents in the order of linear ethers ($E_s < -2$ eV), cyclic carbonates (-1.80 eV $< E_s < -1.75$ eV), linear carbonates (-1.55 eV $< E_s < -1.44$ eV) and cyclic ethers (-1.41 eV $< E_s < -1.20$ eV). In this case, linear ethers exhibit exceptional solvation strength because multiple oxygen ions are coordinated with Na. As solvation number increases and the coordination number of Na is saturated to the maximum value (4 ~ 6 depending on the solvent species), Na ion is significantly stabilized ($E_s < -3.7$ eV) regardless of the solvent species because of the increased number of Na–O coordination. At first glance, these results indicate that most of the Na-solvent complexes with high solvation number could be reversibly intercalated into graphite because of their high E_s. However, the solvents other than linear ethers could not lead to the reversible co-intercalation. In addition, characterizations on ternary GICs composed of Na, linear ethers and graphite suggest that the co-intercalated species have a composition of [Na-solvent], not [Na-solvent$_2$] or [Na-solvent$_3$] which have higher E_s than [Na-solvent] [13]. One of the possible explanations on this behavior is that the desolvation process is involved during the co-intercalation and the highly coordinated complexes are partially desolvated before intercalation. For instance, $[\text{Na-DME}_3]^+$ complex sequentially loses contact with

DME at the expense of the desolvation energy during the co-intercalation process, and finally $[Na\text{-}DME]^+$ intercalates into graphite (Fig. 3.8b). In this scenario, the energy required for each desolvation step determines the final Na-solvent complex candidates for co-intercalation.

The desolvation energies E_{des} of Na-solvent complexes tabulated in Fig. 3.8c are calculated by following equation:

$$E_{des,x} = E_{[Na-solvent_{x-1}]^+} + E_{solvent} - E_{[Na-solvent_x]^+}$$

For all solvent species, $E_{des,1}$ is higher than $E_{des,2}$, which means that the desolvation of one solvent molecule from $[Na\text{-}solvent]^+$ requires more energy than $[Na\text{-}solvent_2]^+$. This is because the interaction of Na and second solvent molecule is shielded by the first solvent molecule, which results in weaker interaction. Based on the experimental reports where the reversible Na-solvent co-intercalation only occurs with linear ethers and the intercalated complexes have the solvation number of 1 [13], we can assume that the complexes with higher solvation number are desolvated during co-intercalation process. According to the desolvation energies in Fig. 3.8c, the highest desolvation energy of $[Na\text{-}solvent_2]^+$ complexes is 1.50 eV. Since $[Na\text{-}solvent_2]^+$ complexes are not observed in graphite galleries, Na-solvent complexes with desolvation energy of less than 1.50 eV are expected to be desolvated before intercalation into graphite. In addition, no electrochemical activity Na-graphite system under DEC (E_s of 1.55 eV) and EC (E_s of 1.75 eV) electrolytes as shown in Fig. 3.6a indicates that the solvation energy of 1.75 eV is not high enough to the co-intercalation. This narrows down the candidates for the co-intercalation as highlighted in Fig. 3.8c.

3.4.3 *Chemical Stability of Na-Solvent Complex*

The thermodynamic stability of Na-solvent complexes is not the sole criteria for the reversible Na-solvent co-intercalation, which is critical for the utilization of graphite as an anode for NIBs. Intercalated complexes should remain stable in graphite galleries because there is a possibility of their decomposition which leads to the cell degradation, for instance, as observed in Li-PC system. A number of researchers have reported that the intercalated Li-PC undergoes the electrochemical decomposition accompanying the evolution of gas phase, leading to the exfoliation of graphite [28, 37, 38]. Therefore, the chemical stability of Na-solvent complexes in graphite galleries should be also examined for the criteria for the co-intercalation.

The chemical stability of a molecule is typically described by energy levels of the highest occupied molecular orbital (HOMO) and the lowest unoccupied molecular orbital (LUMO). For instance, a molecule with high LUMO is stable to reduction reaction because electron should have high energy to occupy the unoccupied molecular orbital of the molecule, i.e., LUMO. To investigate the relative stability of the intercalated $[Na\text{-}solvent]^+$ complexes, we calculated the LUMO of

[Na-solvent]$^+$ molecules. HOMO levels are not considered because the additional electron extraction from the electropositive [Na-solvent]$^+$ molecule is not likely to occur. Figure 3.8d shows the calculated LUMO levels of the candidates for the Na-solvent co-intercalation. LUMO levels of the [Na-linear-ether]$^+$ complexes are relatively higher than the other, although the [Na-DME]$^+$ has comparable value of LUMO. This indicates that it is difficult for electrons to be injected into the [Na-linear-ether]$^+$ complexes, which lowers the possibility of the decomposition reaction and the following degradation of graphite. The [Na-PC]$^+$ complex exhibits the lowest LUMO level which is substantially lower than the Fermi level of graphite, implying more facile chemical reduction upon co-intercalation. SEM image of Na-graphite system in PC electrolyte also supports our calculation (Fig. 3.7). Several cleavages are generated in the sample, which are attributed to the exfoliation of graphite from the decomposition of the intercalated complexes [28]. Thus, only [Na-linear-ether]$^+$ complexes could be reversibly intercalated into graphite among our investigated systems, whereas others could not.

3.4.4 Unified Picture of Na-Solvent Co-intercalation Behavior

The investigation on the Na-solvent co-intercalation behavior and the thermodynamic and chemical stability of Na-solvent complexes is summarized in Fig. 3.9. Our results imply that only [Na-linear-ether]$^+$ complexes could be reversibly intercalated into

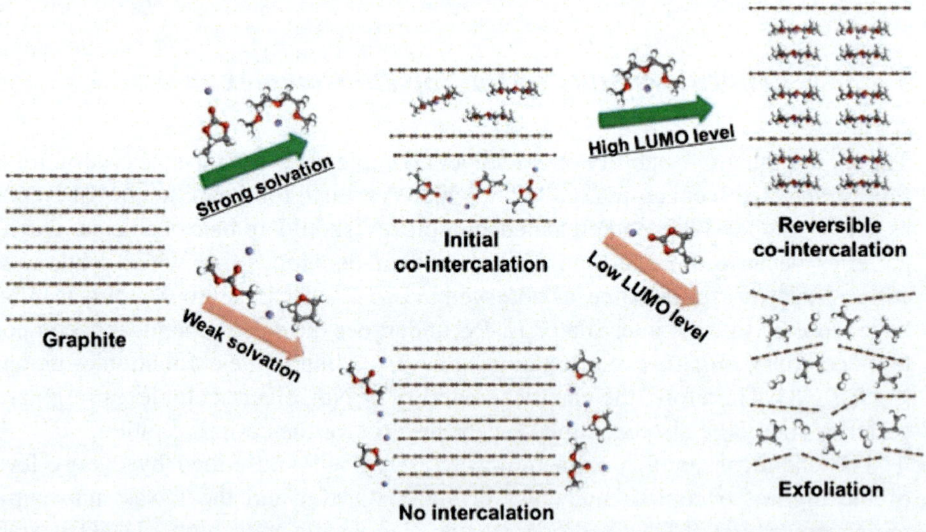

Fig. 3.9 Summary of solvent dependency of the Na-solvent co-intercalation behavior. The structure of Na-solvent co-intercalated graphite (upper right) is adopted from Ref. [13]. Reproduced with permission from WILEY-VCH

graphite because of their high Na solvation energy and chemical stability. If the solvation energy is too low, Na-solvent complexes desolvate before the intercalation, resulting in no intercalation reaction because Na alone could not form GIC. Chemical stability of intercalants is also crucial to prevent the decomposition of intercalated Na-solvent complexes and to ensure the reversible co-intercalation. Our investigation on various groups of solvent species (Fig. 3.6b) supports this claim, as the [Na-linear-ether]$^+$ complexes, which have the highest Na solvation energy and chemical stability are only observed to be electrochemically active among all other complexes.

3.5 Conclusions

We investigated the energy components of AM-GICs (AM = Li, Na, K, Rb, Cs) to understand the thermodynamic instability of Na intercalation into graphite, which prohibits the utilization of graphite as an anode for NIBs. In order to find out the factor determining this phenomenon, we categorized the energy components which contribute to the overall formation energy of AM-GICs. Among the three components, which are the alkali metal destabilization energy, graphite destabilization energy and the interaction energy between alkali metal and graphene layer, our calculations indicated that the peculiarly weak Na-graphene interaction is attributed to the reason for the unstable Na-GIC formation. To utilize the graphite as an anode for NIBs, this unstable Na-graphene interaction should be overcome, and the screening of Na with other molecules such as solvent was proven to be an effective strategy, because it prevents the direct interaction between Na and graphene. We further investigated on the solvent dependency of reversible Na-solvent co-intercalation phenomenon using a various category of solvents such as linear/cyclic ethers and carbonates. It is suggested that the high Na solvation energy and the chemical stability of Na-solvent complexes is critical for the reversible co-intercalation and the utilization of graphite as an anode for NIBs. This finding is further supported by the comparison between the experimental results on the electrochemical activity and the calculated thermodynamic and chemical stability of Na-solvent complexes. We believe that the electrochemical intercalation behavior could be also modulated between co-intercalation and guest ion intercalation in other electrochemical systems by choosing the solvents with desirable solvation energy.

References

1. D. Kundu, E. Talaie, V. Duffort, L.F. Nazar, The emerging chemistry of sodium ion batteries for electrochemical energy storage. Angew. Chem. Int. Ed. **54**, 3431–3448 (2015)
2. N. Yabuuchi, K. Kubota, M. Dahbi, S. Komaba, Research development on sodium-ion batteries. Chem. Rev. **114**, 11636–11682 (2014)
3. M.D. Slater, D. Kim, E. Lee, C.S. Johnson, Sodium-ion batteries. Adv. Funct. Mater. **23**, 947–958 (2013)

4. S.-W. Kim, D.-H. Seo, X. Ma, G. Ceder, K. Kang, Electrode materials for rechargeable sodium-ion batteries: potential alternatives to current lithium-ion batteries. Adv. Energy Mater. **2**, 710–721 (2012)
5. P. Ge, M. Fouletier, Electrochemical intercalation of sodium in graphite. Solid State Ionics **28**, 1172–1175 (1988)
6. D.A. Stevens, J.R. Dahn, The mechanisms of lithium and sodium insertion in carbon materials. J. Electrochem. Soc. **148**, A803–A811 (2001)
7. K. Nobuhara, H. Nakayama, M. Nose, S. Nakanishi, H. Iba, First-principles study of alkali metal-graphite intercalation compounds. J. Power Sources **243**, 585–587 (2013)
8. Y. Okamoto, Density functional theory calculations of alkali metal (Li, Na, and K) graphite intercalation compounds. J. Phys. Chem. C **118**, 16–19 (2014)
9. Z. Wang, S.M. Selbach, T. Grande, Van der Waals density functional study of the energetics of alkali metal intercalation in graphite. RSC Adv. **4**, 4069–4079 (2014)
10. Y. Liu, B.V. Merinov, W.A. Goddard, Origin of low sodium capacity in graphite and generally weak substrate binding of Na and Mg among alkali and alkaline earth metals. Proc. Natl. Acad. Sci. **113**, 3735–3739 (2016)
11. B. Jache, P. Adelhelm, Use of graphite as a highly reversible electrode with superior cycle life for sodium-ion batteries by making use of co-intercalation phenomena. Angew. Chem. Int. Ed. **53**, 10169–10173 (2014)
12. H. Kim, J. Hong, Y.-U. Park, J. Kim, I. Hwang, K. Kang, Sodium storage behavior in natural graphite using ether-based electrolyte systems. Adv. Funct. Mater. **25**, 534–541 (2015)
13. H. Kim, J. Hong, G. Yoon, H. Kim, K.-Y. Park, M.-S. Park, W.-S. Yoon, K. Kang, Sodium intercalation chemistry in graphite. Energy Environ. Sci. **8**, 2963–2969 (2015)
14. Z. Zhu, F. Cheng, Z. Hu, Z. Niu, J. Chen, Highly stable and ultrafast electrode reaction of graphite for sodium ion batteries. J. Power Sources **293**, 626–634 (2015)
15. B. Jache, J.O. Binder, T. Abe, P. Adelhelm, A comparative study on the impact of different glymes and their derivatives as electrolyte solvents for graphite co-intercalation electrodes in lithium-ion and sodium-ion batteries. Phys. Chem. Chem. Phys. **18**, 14299–14316 (2016)
16. G. Kresse, J. Furthmüller, Efficiency of ab-initio total energy calculations for metals and semiconductors using a plane-wave basis set. Comput. Mater. Sci. **6**, 15–50 (1996)
17. J.P. Perdew, K. Burke, M. Ernzerhof, Generalized gradient approximation made simple. Phys. Rev. Lett. **77**, 3865–3868 (1996)
18. P.E. Blöchl, Projector augmented-wave method. Phys. Rev. B **50**, 17953–17979 (1994)
19. G. Kresse, D. Joubert, From ultrasoft pseudopotentials to the projector augmented-wave method. Phys. Rev. B **59**, 1758–1775 (1999)
20. S. Grimme, J. Antony, S. Ehrlich, H. Krieg, A consistent and accurate ab initio parametrization of density functional dispersion correction (DFT-D) for the 94 elements H-Pu. J. Chem. Phys. **132**, 154104 (2010)
21. M.J. Frisch, G.W. Trucks, H.B. Schlegel, G.E. Scuseria, M.A. Robb, J.R. Cheeseman, G. Scalmani, V. Barone, B. Mennucci, G.A. Petersson, H. Nakatsuji, M. Caricato, X. Li, H.P. Hratchian, A.F. Izmaylov, J. Bloino, G. Zheng, J.L. Sonnenberg, M. Hada, M. Ehara, K. Toyota, R. Fukuda, J. Hasegawa, M. Ishida, T. Nakajima, Y. Honda, O. Kitao, H. Nakai, T. Vreven, J.A. Montgomery Jr., J.E. Peralta, F. Ogliaro, M.J. Bearpark, J. Heyd, E.N. Brothers, K.N. Kudin, V.N. Staroverov, R. Kobayashi, J. Normand, K. Raghavachari, A.P. Rendell, J.C. Burant, S.S. Iyengar, J. Tomasi, M. Cossi, N. Rega, N.J. Millam, M. Klene, J.E. Knox, J.B. Cross, V. Bakken, C. Adamo, J. Jaramillo, R. Gomperts, R.E. Stratmann, O. Yazyev, A.J. Austin, R. Cammi, C. Pomelli, J.W. Ochterski, R.L. Martin, K. Morokuma, V.G. Zakrzewski, G.A. Voth, P. Salvador, J.J. Dannenberg, S. Dapprich, A.D. Daniels, Ö. Farkas, J.B. Foresman, J.V. Ortiz, J. Cioslowski, D.J. Fox, *Gaussian 09* (Gaussian, Inc., Wallingford, CT, 2009)
22. A.D. Becke, Density-functional thermochemistry. III. The role of exact exchange. J. Chem. Phys. **98**, 5648–5652 (1993)
23. A.D. McLean, G.S. Chandler, Contracted Gaussian basis sets for molecular calculations. I. Second row atoms, $Z = 11$–18. J. Chem. Phys. **72**, 5639–5648 (1980)

24. D. Guerard, A. Herold, Intercalation of lithium into graphite and other carbons. Carbon **13**, 337–345 (1975)
25. M.S. Dresselhaus, G. Dresselhaus, Intercalation compounds of graphite. Adv. Phys. **51**, 1–186 (2002)
26. W. Rüdorff, E. Schulze, Über Alkaligraphitverbindungen. Z. Anorg. Allg. Chem. **277**, 156–171 (1954)
27. K. Rytkönen, J. Akola, M. Manninen, Density functional study of alkali-metal atoms and monolayers on graphite (0001). Phys. Rev. B **75**, 075401 (2007)
28. H. Buqa, A. Würsig, D. Goers, L.J. Hardwick, M. Holzapfel, P. Novák, F. Krumeich, M.E. Spahr, Behaviour of highly crystalline graphites in lithium-ion cells with propylene carbonate containing electrolytes. J. Power Sources **146**, 134–141 (2005)
29. H. Moon, R. Tatara, T. Mandai, K. Ueno, K. Yoshida, N. Tachikawa, T. Yasuda, K. Dokko, M. Watanabe, Mechanism of Li ion desolvation at the interface of graphite electrode and glyme–Li salt solvate ionic liquids. J. Phys. Chem. C **118**, 20246–20256 (2014)
30. Y. Yamada, Y. Takazawa, K. Miyazaki, T. Abe, Electrochemical lithium intercalation into graphite in dimethyl sulfoxide-based electrolytes: effect of solvation structure of lithium ion. J. Phys. Chem. C **114**, 11680–11685 (2010)
31. K. Xu, A. von Cresce, Interfacing electrolytes with electrodes in Li ion batteries. J. Mater. Chem. **21**, 9849–9864 (2011)
32. M. Inagaki, O. Tanaike, Determining factors for the intercalation into carbon materials from organic solutions. Carbon **39**, 1083–1090 (2001)
33. T. Abe, Y. Mizutani, T. Tabuchi, K. Ikeda, M. Asano, T. Harada, M. Inaba, Z. Ogumi, Intercalation of lithium into natural graphite flakes and heat-treated polyimide films in ether-type solvents by chemical method. J. Power Sources **68**, 216–220 (1997)
34. T. Matsui, K. Takeyama, Li^+ adsorption on a metal electrode from glymes. Electrochim. Acta **43**, 1355–1360 (1998)
35. J.-C. Soetens, C. Millot, B. Maigret, Molecular dynamics simulation of $Li^+BF_4^-$ in ethylene carbonate, propylene carbonate, and dimethyl carbonate solvents. J. Phys. Chem. A **102**, 1055–1061 (1998)
36. W.J. Glover, R.E. Larsen, B.J. Schwartz, Nature of sodium atoms/(Na^+, e^-) contact pairs in liquid tetrahydrofuran. J. Phys. Chem. B **114**, 11535–11543 (2010)
37. M.R. Wagner, J.H. Albering, K.C. Moeller, J.O. Besenhard, M. Winter, XRD evidence for the electrochemical formation of in PC-based electrolytes. Electrochem. Commun. **7**, 947–952 (2005)
38. K. Xu, Nonaqueous liquid electrolytes for lithium-based rechargeable batteries. Chem. Rev. **104**, 4303–4418 (2004)

Chapter 4
Electrochemical Deposition and Stripping Behavior of Li Metal

4.1 Introduction

Recent worldwide drive to the ecofriendly and sustainable life accompanies the increased attention toward the advanced energy storage system. Among various energy storage systems, Li metal batteries can provide high energy density which is sufficient to the application in electric vehicles, due to the low redox potential of Li^+/Li redox couple (-3.04 V vs. standard hydrogen electrode) and the large theoretical capacity (~3860 mAh g^{-1}) of Li metal anode [1, 2]. However, practical application of Li metal anode is still far from realization due to the unreliable and hazardous operation arising from the dendritic growth of Li metal during electrochemical cycling [3–7]. For instance, irregularly grown Li filaments can possibly penetrate the separator and contact with the positive electrode during battery charging, resulting in the short circuit and thermal runaway. In addition, repeating formation and removal of Li dendrites inevitably reveals the fresh surface of Li, which constantly consumes the electrolyte to form solid electrolyte interphase (SEI) layer at the surface of Li. It can also accompany the detachment of small pieces of Li, which is the loss of active material. These undesirable phenomena constantly deteriorate the Coulombic efficiency.

Several strategies have been practiced to meet this challenge. One representative approach is the introduction of mechanically robust layer (e.g. tough and crystalline SEI, artificial protection layer) at the surface of Li metal electrode to physically suppress the reckless growth of Li dendrites [8–17]. For instance, fluoroethylene carbonate (FEC) was adapted as an electrolyte additive to induce the formation of

The content of this chapter has been published in *Chemistry of Materials*. Adapted with permission from [Yoon, G. et al., *Chemistry of Materials* **2018**, 30, 6769]. Copyright (2018) American Chemical Society.

G. Yoon, *Theoretical Study on Graphite and Lithium Metal as Anode Materials for Next-Generation Rechargeable Batteries*, Springer Theses,
https://doi.org/10.1007/978-981-13-8914-6_4

dense and mechanically robust LiF as a major component of SEI layer [10, 11]. Physical coating of protection layers such as Al_2O_3 and carbon has also been proven effective in many researches [14–17]. These mechanical treatments were partly successful in enhancing the stability of Li metal anode, as confirmed by the prolonged cycle life despite eventual failure. On the other hand, chemical strategies have been employed to control the behavior of Li ion near the surface, since the propagation of dendrites is accelerated by highly localized Li ion concentration at the initial protrusion [18–23]. Coating conductive and uniform SEI layers by using SO_2 gas additive [18] or adding conductive inorganics such as $LiNO_3$ [20], LiBr [22] and lithium oxalyldifluoroborate (LiODFB) [23] into electrolyte could facilitate the Li ion movement at the surface, well distributing the ion flux at the surface and relieving the severe localization. Introducing Cs ion, which does not reduce to Cs metal in cell operating voltages, was also proven effective in distributing Li ions by electrostatic repulsion between Li ions and Cs ions [24]. Despite a notable improvement on the electrochemical performance by these intensive efforts, the fundamental inhibition of dendritic growth at the initial stage of deposition has not yet been realized. Thus, most of the strategies still focus on the treatment of grown dendrites, not on the suppression of initial onset of growth.

There have been theoretical approaches to understand the fundamentals on the initial Li deposition behavior. Early works by Rosso and Chazalviel et al. investigated the initial nucleation and growth behavior of Li dendrite using analytical model of ion concentrations in the system [25–28]. Through the understanding on the evolution of ion concentrations in a simplified uniform one-dimensional model, they proved that the ion concentration near the electrode is either converged into a fixed value or falls into zero after a certain time. The time at the depletion is called Sand's time [29], which is proportional to the ion diffusivity at the electrolyte, initial ion concentration and inversely proportional to the current density and anionic transport number. After the depletion, the potential starts diverging and the deposition occurs dendritic in order for the system to escape the instability [27]. Recently, the change of growth mechanism is directly visualized by optical microscope, where the mossy growth is observed before Sand's time and the dendritic growth is observed after Sand's time [30]. These works have been successful in describing the initial growth behavior of Li deposits. However, there is still a room for a systematic understanding on the reversible Li electrodeposition and stripping behavior with respect to the experimentally controllable conditions, which potentially helps practical application of Li metal batteries.

In this work, we systematically control the experimental conditions and demonstrate their implications to the Li deposition behavior by performing continuum mechanics simulations. The rate of Li deposition, the shape of electrode surface, the conductivity and uniformity of SEI layer and the repeating cycles of Li deposition and stripping were taken as representative factors affecting the geometry of Li growth. Our simulations not only demonstrate the implications of deposition rate, SEI layers and surface geometries to Li growth behavior, but also provide a hint at the origin of dead Li formation and different morphology evolutions of Li deposits.

We believe that this study provides an insight to designing a stable Li metal anode to realize the safe and stable operation of Li metal batteries.

4.2 Computational Details

All simulations on lithium electrodeposition were conducted using electrodeposition module with tertiary current distribution, Nernst–Planck interface in COMSOL multiphysics [31]. Behaviors of charged species in an electrolyte were treated with Nernst–Planck equation with an assumption of charge conservation as described below.

$$\boldsymbol{N}_i = -\boldsymbol{D}_i \nabla c_i - z_i u_i F c_i \nabla \varphi_i$$

$$\sum_i z_i c_i = 0.$$

Here, $\boldsymbol{N}_i$ represents the flux, $\boldsymbol{D}_i$ the diffusion coefficient, c_i the concentration, z_i the valence, u_i the mobility, φ_i the electrolyte potential of each species and F a Faraday constant. Contribution of convection to a flux was neglected. The material balances were conserved using the following mass conservation law:

$$\frac{\partial c_i}{\partial t} + \nabla \cdot \boldsymbol{N}_i = 0.$$

The reaction kinetics at the electrode was described by concentration dependent Butler–Volmer equation

$$i = i_0\left(C_R \exp\left(\frac{\alpha_a F \eta}{RT}\right) - C_O\left(\frac{-\alpha_c F \eta}{RT}\right)\right)$$

where i_0 is the exchange current density, C_R, C_O the dimensionless concentrations of reduced and oxidized species, α_a, α_c the anodic and cathodic charge transfer coefficient, η the activation overpotential, R the gas constant and T the temperature. The resulting electrodeposition was assumed to occur in the normal direction to the boundary with a velocity v:

$$v = \frac{i}{nF}\frac{M}{\rho},$$

where M and ρ are the molar mass and density of Li, respectively.

The equations above govern the ion transport behavior in the electrolyte and the electrodeposition reaction at the electrode/electrolyte surface. All the other boundaries were treated as insulating, as described by the following equation:

$$\boldsymbol{N}_i \cdot \boldsymbol{n} = 0.$$

Here, $\boldsymbol{n}$ represents the vector normal to each boundary.

Each charge transfer coefficient was set to 0.5 and temperature was fixed at 300 K in all simulations. Initial Li ion diffusion coefficient in the electrolyte is set to $7.5 \times 10^{-11}\ m^2\ s^{-1}$. We used initial Li ion concentration of 1 M and the overpotential for Li deposition is controlled from 0.1 to 0.4 V to control the rate of deposition and stripping reaction. Other input parameters were properly controlled to see the effect of variables, such as the rate of Li deposition, shape of electrode surface, SEI layer conductivity and uniformity, and the repeating cycle of Li deposition and stripping.

The growth behavior of Li deposits has been observed in several previous studies, which revealed that the shape of Li deposits is highly sensitive to the experimental conditions [3–7]. In order to systematically understand the effect of each variable, we take the rate of Li deposition, the shape of electrode surface, the conductivity and uniformity of SEI layer and the repeating cycles of Li deposition and stripping as important factors affecting the Li deposition behavior and theoretically investigate their implications.

4.3 Effect of Deposition Rate

The standard model used for overall simulation is described in Fig. 4.1a. Since the reaction on the flat and clean surface without any irregularity necessarily results in uniform and monotonous deposition in continuum mechanics simulation, we intentionally built a surface with bumps to induce the irregularity of ion flux and try to reflect the real experimental situation. Using this model as starting geometry, Li deposition behavior at the initial stage was simulated with different rates.

First, initial deposition under relatively slow rate was simulated and described in Fig. 4.1b. Here, background color indicates the Li ion concentration in the electrolyte, green lines are equipotential lines, and red arrows represent the intensity and the direction of electrolyte current density. It is observed that early deposition was focused on the corners of surface bumps, in agreement with previous understanding. If we take a closer look at the initial stage of deposition (see Fig. 4.2 for time-dependent geometry evolution), it is shown that equipotential lines and current densities are concentrated around the corners, which makes those corners preferential deposition sites, or deposition seeds. In addition, the surface Li ion concentration is converged to non-zero value, indicating that this condition does not induce the complete depletion of Li ion at the surface. We also notice that the growth direction of Li deposits is less directional under this condition, not unidirectional or dendritic with branches.

On the other hand, high-rate deposition with identical Li utilization yields vertical Li growth along with branches, as described in Fig. 4.1c. Equipotential lines displayed in Fig. 4.1c are significantly denser than the ones in Fig. 4.1b due to the large overpotential condition, meaning the magnitude of electrolyte current density, which

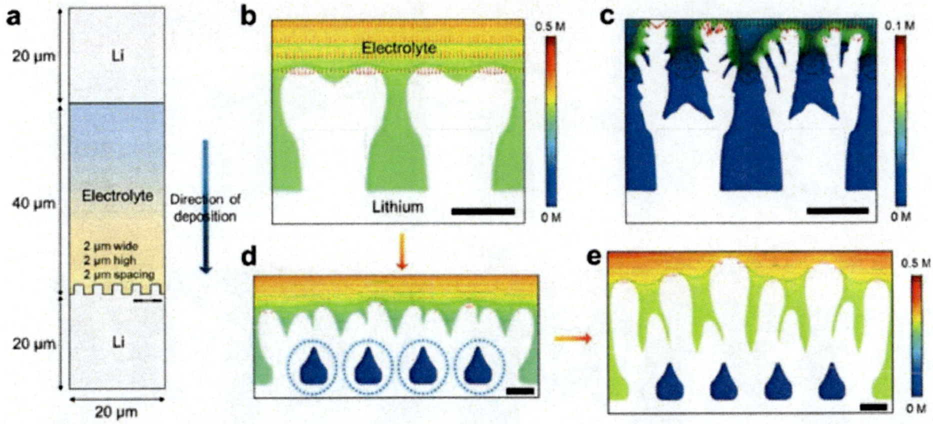

Fig. 4.1 Evolution of Li deposits for different deposition rates. **a** Reference model used for simulations. Geometry after initial deposition at **b** slow and **c** fast deposition rates. The gray lines represent the initial geometry before deposition. **d** Expected shape after continuous deposition at slow rate. The dotted circles represent pores that could be generated after physical contact between deposits. **e** Geometry after further deposition from (**d**). Deposits from neighboring bumps will meet and eventually form a mossy-like shape, whereas the high-rate condition yields vertical growth of Li with many branches, forming a needle-like shape. The scale bars are 2-μm long. The utilization of Li is identical in (**b**) and (**c**). The contour levels of the potential are displayed every 2.5 meV. Reproduced with permission from American Chemical Society

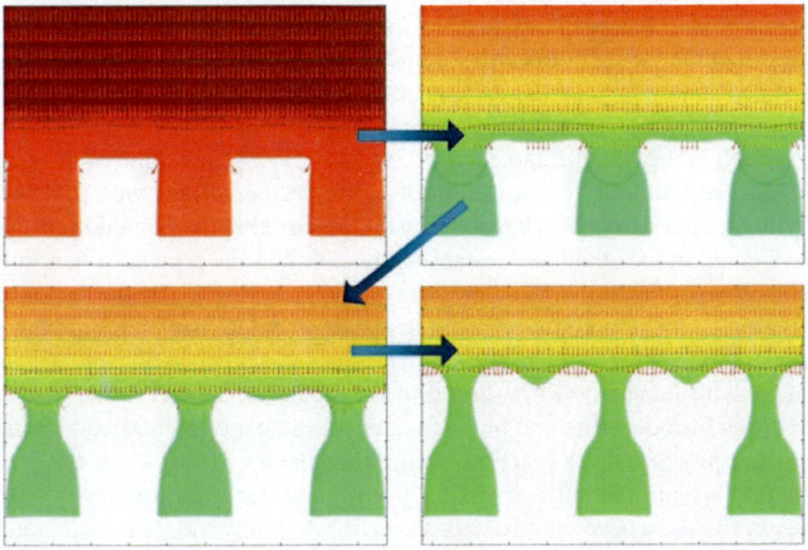

Fig. 4.2 Snapshots for time-dependent Li electrodeposition using a relatively slow rate. Reproduced with permission from American Chemical Society

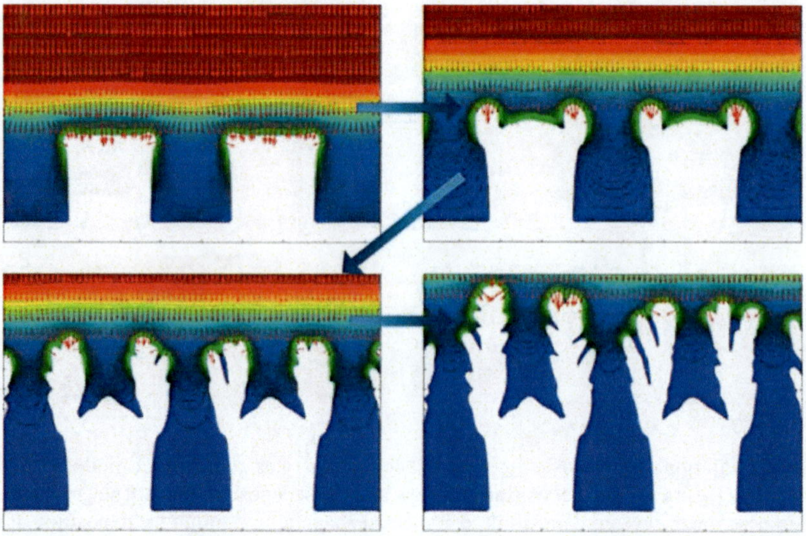

Fig. 4.3 Snapshots for time-dependent Li electrodeposition using a relatively fast rate. Reproduced with permission from American Chemical Society

is proportional to the gradient of potential, is larger in high-rate condition. Because of this large current density, even a little inhomogeneity after an initial deposition at the corners would induce a non-negligible disturbance of local current densities, leading to the directional growth towards counter electrode and the formation of branches (Fig. 4.3). Another significant difference from low-rate condition is in the depletion of Li ion concentration at the surface arising from this large current density, which is described in the background color of Fig. 4.1c. The observation of directional and branching growth with the depletion of Li ion at the surface well agrees with the early theoretical works of Chazalviel, which claim that the depletion of ions at the surface could induce the local instability, resulting in the growth of dendrites [27].

Since the initial growth under slow deposition (Fig. 4.1b) is less directional, continuing deposition could possibly result in a contact between growing deposits from neighboring corners. Figure 4.1d shows the expected geometry after the contacts, with inevitable formation of pores upon the contact. Note that the irregularity was introduced to the bumps to reflect the experimental conditions. When we further proceeded the reaction from the geometry in Fig. 4.1d, the preferential deposition around the surface arising from the surface irregularity is observed as shown in Fig. 4.1e. Growing Li deposits will eventually meet each other again, and pores will be generated as the deposits aggregate. Repeating process of this contact and pore generation would lead to the porous geometry, which resembles the mossy shape observed in experiments [30].

Taking these simulations together, it is summarized that the initial deposition is focused at deposition seeds where equipotential lines and current densities converge, and preferential deposition during the growth is more noticeable with high-rate condition. The current densities from slow deposition condition is not sufficient enough to induce preferential deposition from minor irregularity, leading to the less directional growth which eventually forms a porous and mossy structure. On the other hand, high-rate condition results in a complete depletion of Li ion at the surface and the preferential deposition even with a little inhomogeneity, leading to the unidirectional and dendritic growth with many branches.

4.4 Effect of Surface Geometry

Since we observed the important role of deposition seeds and the local irregularity in the growth behavior of Li in the previous section, we attempted to understand the effect of surface geometry by simulating Li deposition reaction using different initial surface shapes. Other than the square-shape bumps used in previous section, triangular- and circular-shaped bumps were considered as extreme cases for surface irregularity, as shown in Fig. 4.4. All other simulation parameters such as initial ion concentration, Li ion diffusivity of electrolyte, Li utilization and rate of deposition are maintained from the conditions used for Fig. 4.1c. Figure 4.5a shows the shape of Li deposits starting from the surface with triangular bumps. Since this initial geometry (Fig. 4.4) has sharp corners exposed to the electrolyte which could act as deposition seeds and the number of corners are reduced to half compared to the model with square bumps, it is natural to expect that the extremely directional and branching growth would be focused on those corners. Indeed, a number of branches were generated from the initial stage of deposition as observed in Fig. 4.6. In addition, the final height of the deposits is larger than that of Fig. 4.1c, indicating more directional growth due to the smaller number of initial deposition seeds (Figs. 4.5 and 4.6).

Fig. 4.4 Models used for simulation with **a** triangular and **b** circular surface shape. Reproduced with permission from American Chemical Society

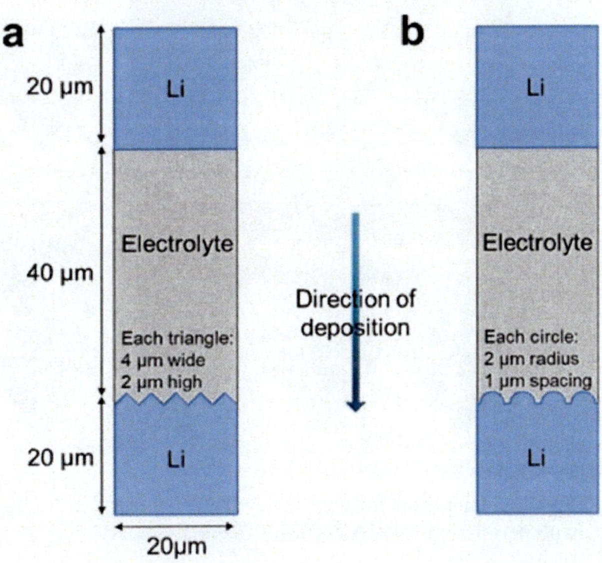

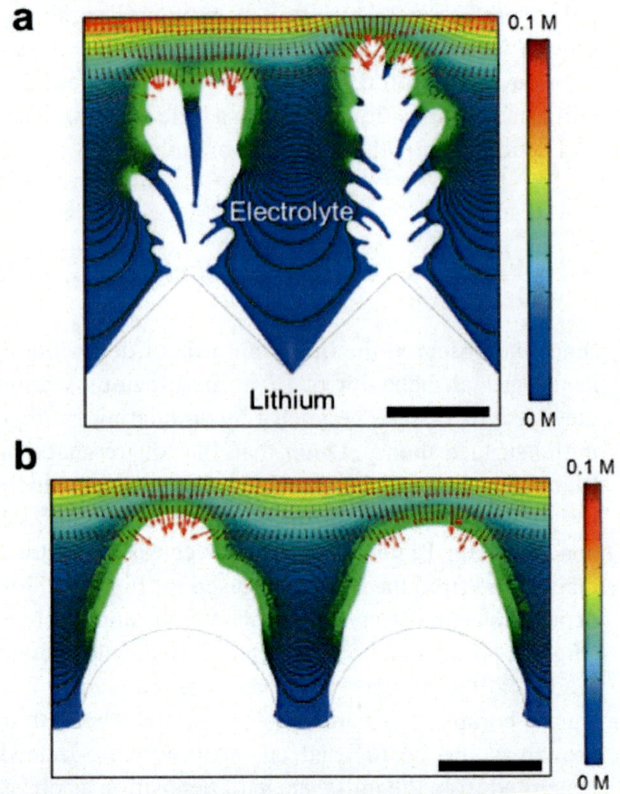

Fig. 4.5 Evolution of Li deposits starting from an initial surface geometry with **a** triangular and **b** circular bumps. The gray lines represent the initial geometry before deposition, and the scale bars are 2-μm long. The utilization of Li is identical in (**a**) and (**b**), and the contour levels of potential are displayed every 2.5 meV. Reproduced with permission from American Chemical Society

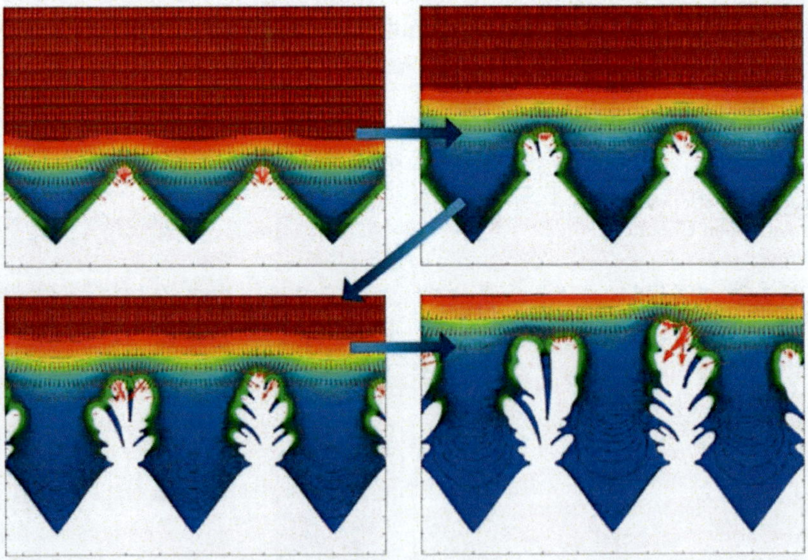

Fig. 4.6 Snapshots for time-dependent Li electrodeposition at the triangular surface. Reproduced with permission from American Chemical Society

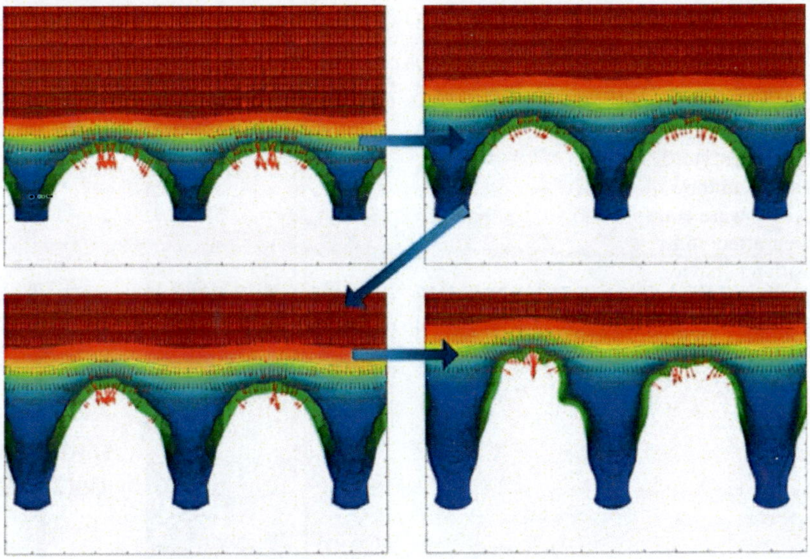

Fig. 4.7 Snapshots for time-dependent Li electrodeposition at the circular surface. Reproduced with permission from American Chemical Society

On the other end of extreme, the deposition on the surface with smooth circular bumps yields quite uniform and dense growth as described in Fig. 4.5b. The lack of irregular spot at the surface led to the well distributed initial total current along the smooth surface with large area, different from the current concentration at certain points as observed in Fig. 4.7. In other words, this model initially has large active surface area where deposition reaction occurs, so the magnitude of local current densities is relatively small. Therefore, smooth surfaces have a larger tolerance against branching when a little irregularity was introduced during the deposition reaction, similar to that observed in Fig. 4.1b, c, where the difference of the magnitude of current densities resulted in conflicting branching behavior. We would like to note that if deposition further proceeds and the surface irregularities are accumulated, there will be a point when the preferential growth occurs with the generation of branches. Still, smoother initial surface delays the onset of branching, giving larger room for a dense and uniform deposition of Li within certain utilization which is crucial for practical application.

4.5 Implications of SEI Layer Properties

In previous sections, we used simple models consisting of Li metal electrodes and electrolyte for electrodeposition simulation. However, it is widely known that SEI layer generated at the surface of electrodes by the decomposition of electrolyte plays a crucial role in battery operation [1, 32]. Therefore, it is imperative to see the effect

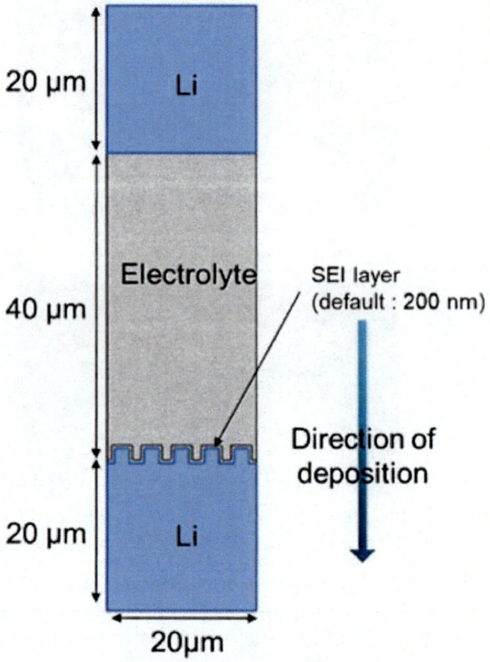

Fig. 4.8 Model used for simulation with 200 nm-thick SEI layers. Li ion diffusivity in SEI layers is 100 times as sluggish as that in electrolyte. The thickness and the Li ion diffusivity were controlled to see their effect in Li deposition behavior. Reproduced with permission from American Chemical Society

of SEI layers and their properties such as Li ion diffusivity on Li growth behavior. Typical SEI layers in practical cells have very low electrical conductivity and high Li ion conductivity, preventing the further electrochemical decomposition of electrolyte while allowing the passage of Li ion through them. Taking consideration of this property of SEI layer, we introduced artificial layer at the surface of electrode, which has 200 nm thickness with 100 times lower Li ion diffusivity than the electrolyte, as shown in Fig. 4.8. Since the electron cannot be provided through SEI layer, the electrodeposition reaction was set to occur only at the surface of the electrode.

Figure 4.9a describes the Li deposition behavior with the presence of SEI layer at the surface of the electrode. The most notable difference comparing to the previous simulations is that Li growth is much more dendritic, indicating that the tendency for preferential deposition is far greater in this condition. It should be also noted that the deposition rate used for the simulation of Figs. 4.1c and 4.5 results in extremely severe branching growth in this case, therefore slower deposition rate was used for simulations in Fig. 4.9 for better observation of Li deposits. To understand the much severe preferential deposition, we investigated deeper on the behavior of local Li concentration. It is found that the Li ion concentration is very high outside the SEI layer and rapidly decreases as approaching the electrode surface within the SEI layer, as described by background color in Fig. 4.9a. Since the movement of Li ion is fast in electrolyte but hampered within sluggish SEI layer, the supply of Li ion to the electrode surface is not fast enough for incessant reaction, so Li ions accumulate at the outside of SEI layer. Li ion concentration along the vertical line is plotted in Fig. 4.9a, which also confirms the abrupt decrease of the concentration

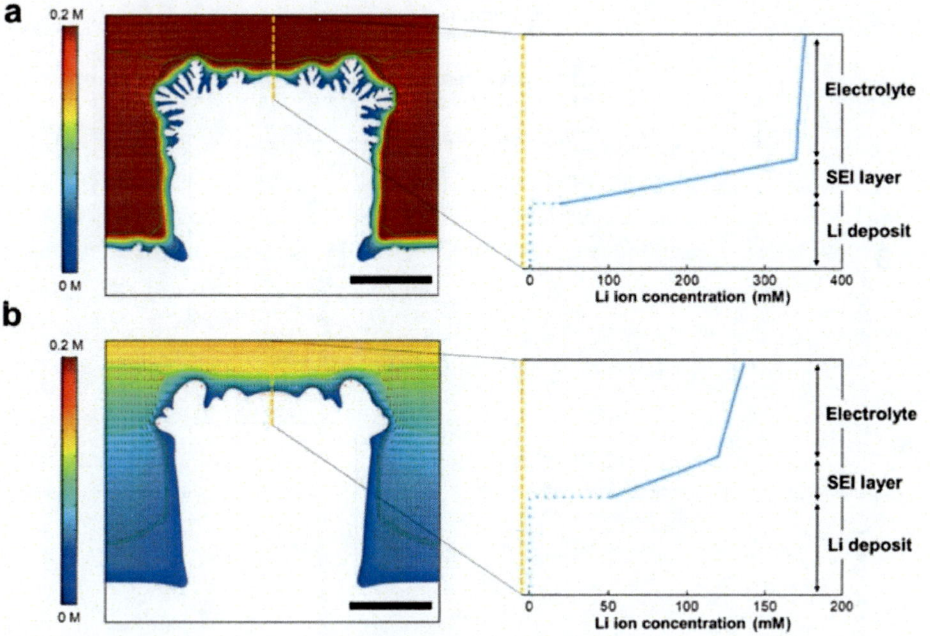

Fig. 4.9 Evolution of Li deposits with the presence of a 200-nm-thick surface SEI layer. The Li ion concentration profile along the yellow dotted line is also shown. The Li ion conductivity of the layer was set to **a** 100 times and **b** 10 times lower than that of the electrolyte. The gray lines represent the initial geometry before deposition, and the scale bars are 1-μm long. The utilization of Li is identical in (**a**) and (**b**), and the contour levels of potential in (**a**) and (**b**) are displayed every 10 meV. Reproduced with permission from American Chemical Society

in SEI layer. The severe concentration gradient in SEI layer results in high local current density, therefore, highly selective deposition can occur with an introduction of minor irregularity during the deposition.

Since the SEI layer in the simulation is described by simply reducing the ionic diffusivity of electrolyte, it retains a bi-ionic conducting property where both Li ion and anion are mobile within SEI layer. However, considering that typical SEI layer is mostly composed of inorganic compounds, it could be treated as single-ionic conductor. In this regard, we conducted a simulation using the SEI layer with single-ionic conducting nature, while maintaining all other condition. Figure 4.10 compares the Li ion and the potential profile in electrolyte region, when single and bi-ionic conducting SEI layer is adopted. Since the anion is immobile in single-ionic conductor, Li ion concentration gradient was not observed in single-ionic conducting SEI as shown in Fig. 4.10a. In addition, we found that the potential gradient is linear, indicating that the Ohmic drop in electrolyte is the sole source affecting the potential profile. On the other hand, the potential gradient is parabolic in bi-ionic conducting SEI, implying that both Ohmic drop and the Li ion concentration gradient determines the potential profile. In spite of this discrepancy, severe branching growth is commonly observed, as described in Fig. 4.11. This is because sluggish SEI acts

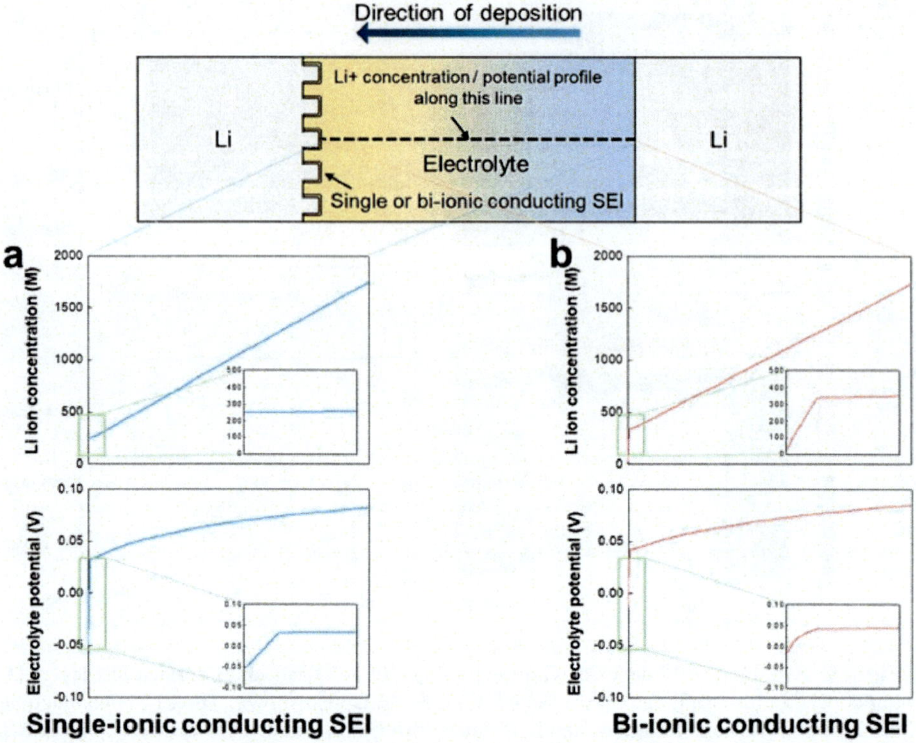

Fig. 4.10 Li ion concentration and electrolyte potential profile in electrolyte region using **a** single-ionic conducting and **b** bi-ionic conducting SEI layer. Reproduced with permission from American Chemical Society

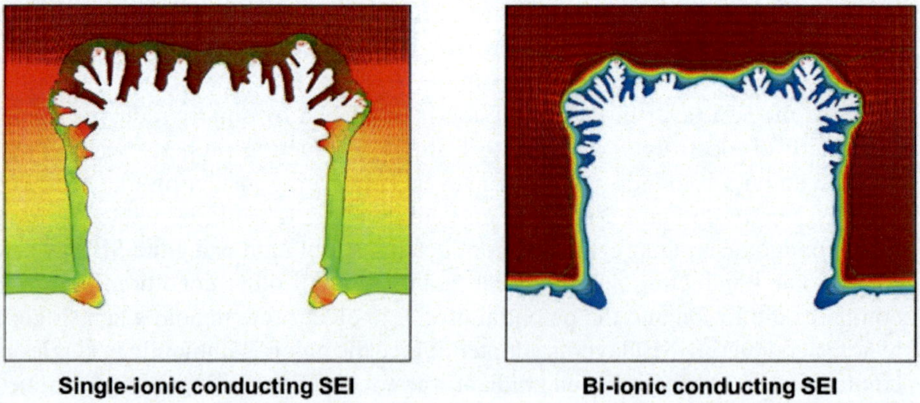

Fig. 4.11 Li deposition shape using single and bi-ionic conducting SEI. Reproduced with permission from American Chemical Society

as high resistance regardless of its conducting nature, resulting in steep Ohmic drop in SEI layer. Therefore, high local current density is induced at the surface, and the generation of minor protrusions leads to the branching growth as we discussed in previous sections.

Our observations hitherto consistently suggest that the high local current density at the surface induced by the steep potential gradient leads to the unwanted branching growth of Li. In this respect, the thickness of the electrolyte could be an important design criteria for uniform Li deposition, since the potential gradient between the anode and the cathode is determined by the thickness of the electrolyte. The impact would be dramatic for solid electrolytes, particularly for the case when the Li metal and solid electrolytes are stabilized without the formation of SEI layer, because the potential gradient is then linear throughout the electrolyte and the electrolyte thickness directly increases the gradient of potential profile. In this case, as described in Fig. 4.12, the Ohmic drop is the sole source determining the potential profile in the electrolyte region. It would result in the linear potential gradient near the solid electrolyte interface without the Li ion concentration gradient, thus much smoother Li metal growth can be achieved (Fig. 4.13). Moreover, it implies that the thickness of

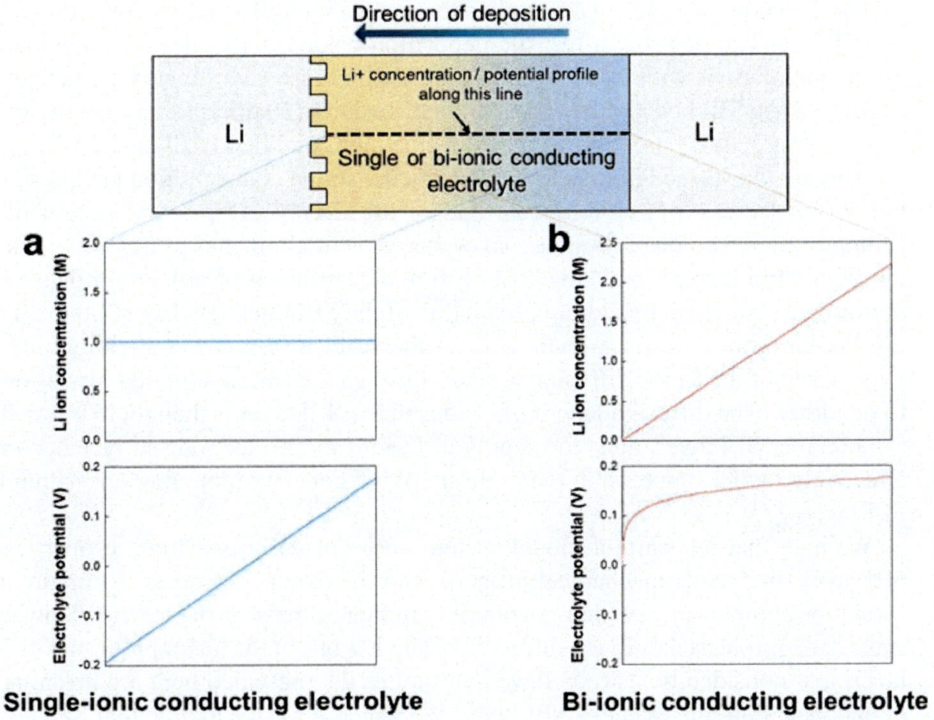

Fig. 4.12 Li ion concentration and electrolyte potential profile in electrolyte region using **a** single-ionic conducting and **b** bi-ionic conducting electrolyte. It is assumed that the interface between the Li metal and the electrolyte is stabilized without the formation of SEI layer. Reproduced with permission from American Chemical Society

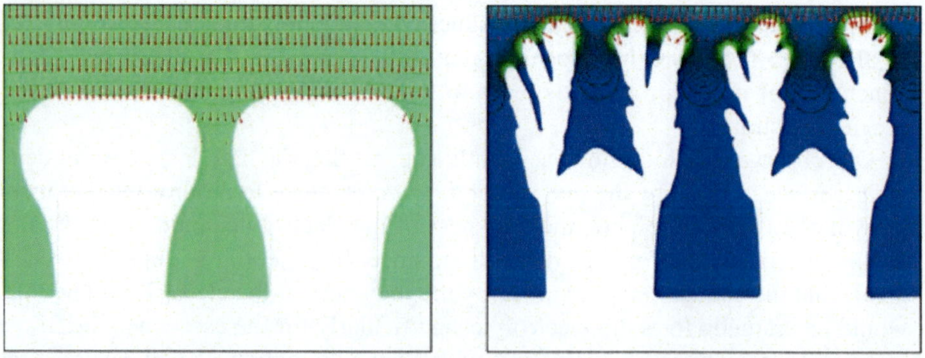

Fig. 4.13 Li deposition shape using single and bi-ionic conducting electrolyte. It is assumed that the interface between the Li metal and the electrolyte is stabilized without the formation of SEI layer. Reproduced with permission from American Chemical Society

the single ionic conductor (or solid electrolyte) could be an important design criteria for the uniform Li deposition, since the linear potential gradient near the surface would be governed by the thickness of the electrolyte with a given applied voltage.

The observation in Fig. 4.9a implies that the Li ion diffusivity and the thickness of SEI layer can possibly affect the deposition behavior, therefore, we conducted further simulations with those factors controlled. Figure 4.9b shows the shape of Li deposits when SEI layer with higher Li ion diffusivity (10 times faster than the value used for Fig. 4.9a) is used. As Li ions become more mobile in SEI layer, the supply of Li ion at the surface of the electrode is facile, so the concentration profile shows less abrupt decrease of Li ion concentration in the SEI layer (Fig. 4.9b). The resulting geometry therefore displays much smoother growth compared to Fig. 4.9a, which implies that inducing conductive SEI layer at the surface is beneficial for dense and uniform growth of Li. In addition, reduction of the SEI thickness from 200 to 50 nm resulted in more uniform growth of Li at the surface (Fig. 4.14), highlighting the critical role of the Li ion diffusion kinetics through the SEI layer in the formation of Li dendrites. The diffusion length of Li ions in the SEI layer is thought to be smaller for a thinner SEI layer; thus, the supply of Li ions to the electrode surface becomes effectively facile, suppressing the build-up of the concentration gradient within the SEI layer.

We note that our current model system does not describe certain properties of SEI layer such as its mechanical property and the dynamic process of rupture and formation at the surface, which can possibly influence the growth behavior. However, since our approach takes Li ion diffusivity, which is one of the major properties of SEI layer, into consideration, we believe that the results presented here are meaningful in understanding the effect of SEI layer. We would also like to note that SEI layers observed in experiments are irregular in terms of thickness and the conductivity. In our recent work, we tried to address the irregularity of SEI layer by mixing the layers with different Li ion diffusivity, and observed more irregular growth of Li when the irregularity of SEI layers is introduced as inferred from simulations in this work [18].

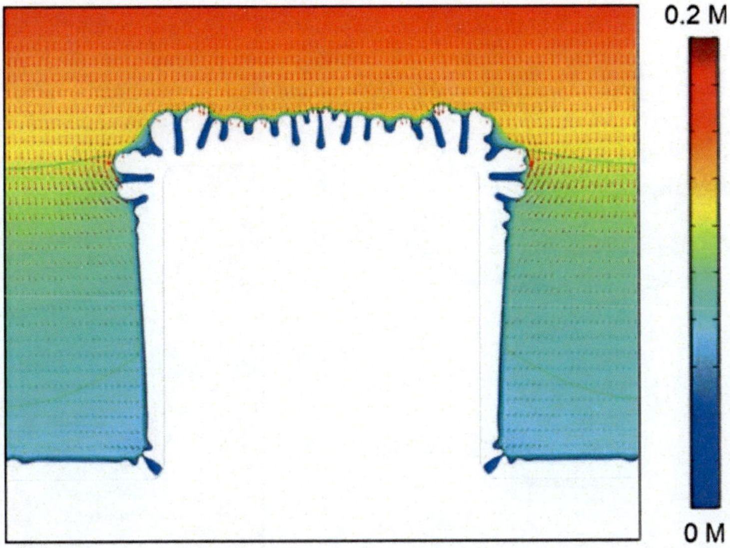

Fig. 4.14 Shape of Li deposits after a deposition simulation with 50 nm-thick SEI layer. Reproduced with permission from American Chemical Society

4.6 Consequences of the History of Deposition and Stripping

Our simulations so far attempted to investigate an effect of various conditions on the first deposition step of Li. However, unlike typical metal electrodepositions, practical batteries operate for thousands of charge and discharge cycles, which means that Li metal anode undergoes a number of repeating deposition and stripping process. Therefore, it is important to see how the history of deposition and stripping affects the Li growth behavior. In particular, we used the geometry obtained after a single deposition step with two different deposition rates as starting structures, and applied reverse bias with different stripping rates to see the morphology evolution during the stripping process.

Figure 4.15 shows the geometry evolution of Li deposit after a single stripping step from fast deposited Li. When we strip with fast rate, the morphology goes back to the initial geometry (Fig. 4.1a) with high reversibility. As our simulations in Fig. 4.1 demonstrated that electrodeposition is preferred to occur at the corners where the equipotential line and the electrolyte current density is concentrated, the stripping reaction is also more active at the corners, or at the points with large curvature. The high selectivity in high rate condition therefore leads to the reduction of curved points, which results in highly reversible stripping. However, this selectivity would be lower in slow stripping condition as described in Fig. 4.1. Uniform stripping around the highly irregular surface means that the reversibility is significantly low in slow stripping condition, and we found that the irreversibility can be particularly observed at the narrow points in the deposits, as shown in Fig. 4.15. Relatively

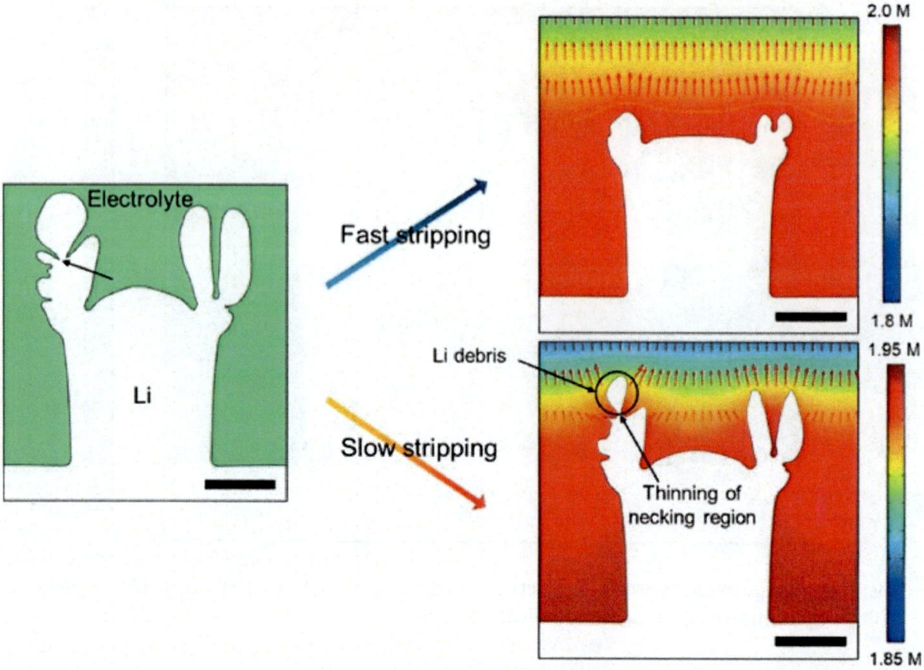

Fig. 4.15 Li stripping behavior from irregular deposit. Dead Li formation is shown at the narrow point (black arrow) for the slow stripping condition. Complete stripping was not achieved because of the failure of the solution converge, presumably arising from the dynamic generation of singular points during the stripping reaction. The scale bars are 1 μm long. Reproduced with permission from American Chemical Society

isotropic stripping will eventually make these bottlenecks extremely thin, resulting in the separation of a fragment of Li deposits from the bulk electrode. To confirm that this thinning at the slow stripping is generally observed at the narrow points, we artificially made a model with those points and performed a stripping simulation (Fig. 4.16), which also displays the similar behavior. This phenomenon called the formation of dead Li has been observed in many experiments, and causes the severe

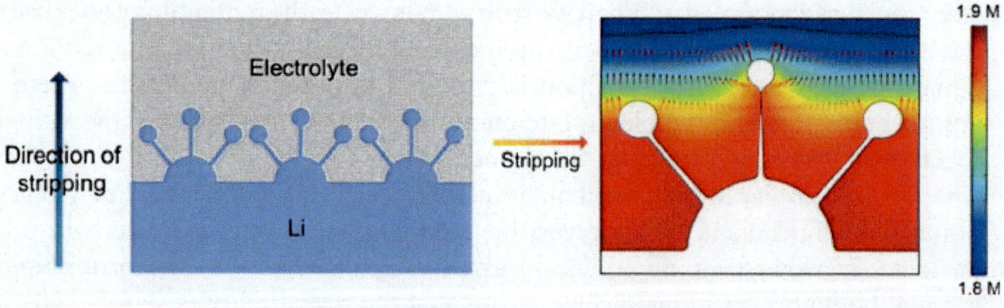

Fig. 4.16 Stripping from artificially made model with narrow points. Reproduced with permission from American Chemical Society

degradation of cycle stability because it is a permanent loss of active material [33, 34]. Our simulations in Figs. 4.15 and 4.16 demonstrate that the stripping rate is more critical than deposition rate to the formation of dead Li, in contrast with the conventional belief that the faster deposition kinetics would lead to the formation of Li debris. Since the stripping of Li metal corresponds to the discharge process whose rate is not controllable, it is imperative to find a way to suppress the formation of dead Li for the commercialization of Li metal anode.

4.7 Conclusions

In this work, we attempted to simulate the Li deposition and stripping behavior using continuum mechanics with the consideration of various experimental conditions such as the reaction rate, surface morphology, Li ion diffusivity and the thickness of SEI layer, and the cycling process of deposition and stripping. Our results can be briefly summarized as (1) preferential Li growth occurs because of the disturbance of the local current density in the presence of irregular surface properties, (2) the tendency of preferential growth determines the initial morphology of Li deposits (mossy or branching), (3) the preference depends on the surface irregularity from the geometric disturbance and SEI properties, and the magnitude of local current density at the surface, and (4) stripping irregular deposits with slow rate induces isotropic reaction, potentially resulting in the formation of dead Li. Although our model could not completely reflect the experimental condition, our findings provide important clues for the mechanism of dendritic growth and the formation of dead Li, and we hope that these discoveries give a hint toward the stable operation of Li metal anode.

References

1. J.M. Tarascon, M. Armand, Issues and challenges facing rechargeable lithium batteries. Nature **414**, 359 (2001)
2. J.W. Choi, D. Aurbach, Promise and reality of post-lithium-ion batteries with high energy densities. Nat. Rev. Mater. **1**, 16013 (2016)
3. Z. Li, J. Huang, B. Yann Liaw, V. Metzler, J. Zhang, A review of lithium deposition in lithium-ion and lithium metal secondary batteries. J. Power Sources **254**, 168–182 (2014)
4. K. Zhang, G.-H. Lee, M. Park, W. Li, Y.-M. Kang, Recent developments of the lithium metal anode for rechargeable non-aqueous batteries. Adv. Energy Mater. **6**, 1600811 (2016)
5. Y. Guo, H. Li, T. Zhai, Reviving lithium-metal anodes for next-generation high-energy batteries. Adv. Mater. **29**, 1700007 (2017)
6. K.N. Wood, M. Noked, N.P. Dasgupta, Lithium metal anodes: toward an improved understanding of coupled morphological, electrochemical, and mechanical behavior. ACS Energy Lett. **2**, 664–672 (2017)
7. W. Xu, J. Wang, F. Ding, X. Chen, E. Nasybulin, Y. Zhang, J.-G. Zhang, Lithium metal anodes for rechargeable batteries. Energy Environ. Sci. **7**, 513–537 (2014)

8. K. Leung, F. Soto, K. Hankins, P.B. Balbuena, K.L. Harrison, Stability of solid electrolyte interphase components on lithium metal and reactive anode material surfaces. J. Phys. Chem. C **120**, 6302–6313 (2016)
9. X. Chen, T.-Z. Hou, B. Li, C. Yan, L. Zhu, C. Guan, X.-B. Cheng, H.-J. Peng, J.-Q. Huang, Q. Zhang, Towards stable lithium-sulfur batteries: mechanistic insights into electrolyte decomposition on lithium metal anode. Energy Storage Mater. **8**, 194–201 (2017)
10. C. Zu, N. Azimi, Z. Zhang, A. Manthiram, Insight into lithium-metal anodes in lithium-sulfur batteries with a fluorinated ether electrolyte. J. Mater. Chem. A **3**, 14864–14870 (2015)
11. R. Miao, J. Yang, Z. Xu, J. Wang, Y. Nuli, L. Sun, A new ether-based electrolyte for dendrite-free lithium-metal based rechargeable batteries. Sci. Rep. **6**, 21771 (2016)
12. C. Monroe, J. Newman, The impact of elastic deformation on deposition kinetics at lithium/polymer interfaces. J. Electrochem. Soc. **152**, A396–A404 (2005)
13. K. Yan, H.-W. Lee, T. Gao, G. Zheng, H. Yao, H. Wang, Z. Lu, Y. Zhou, Z. Liang, Z. Liu, S. Chu, Y. Cui, Ultrathin two-dimensional atomic crystals as stable interfacial layer for improvement of lithium metal anode. Nano Lett. **14**, 6016–6022 (2014)
14. E. Kazyak, K.N. Wood, N.P. Dasgupta, Improved cycle life and stability of lithium metal anodes through ultrathin atomic layer deposition surface treatments. Chem. Mater. **27**, 6457–6462 (2015)
15. A.C. Kozen, C.-F. Lin, A.J. Pearse, M.A. Schroeder, X. Han, L. Hu, S.-B. Lee, G.W. Rubloff, M. Noked, Next-generation lithium metal anode engineering via atomic layer deposition. ACS Nano **9**, 5884–5892 (2015)
16. G. Zheng, S.W. Lee, Z. Liang, H.-W. Lee, K. Yan, H. Yao, H. Wang, W. Li, S. Chu, Y. Cui, Interconnected hollow carbon nanospheres for stable lithium metal anodes. Nat. Nanotechnol. **9**, 618 (2014)
17. H. Lee, D.J. Lee, Y.-J. Kim, J.-K. Park, H.-T. Kim, A simple composite protective layer coating that enhances the cycling stability of lithium metal batteries. J. Power Sources **284**, 103–108 (2015)
18. S. Moon, H. Park, G. Yoon, M.H. Lee, K.-Y. Park, K. Kang, Simple and effective gas-phase doping for lithium metal protection in lithium metal batteries. Chem. Mater. **29**, 9182–9191 (2017)
19. D. Aurbach, Review of selected electrode–solution interactions which determine the performance of Li and Li ion batteries. J. Power Sources **89**, 206–218 (2000)
20. J. Guo, Z. Wen, M. Wu, J. Jin, Y. Liu, Vinylene carbonate–$LiNO_3$: a hybrid additive in carbonic ester electrolytes for SEI modification on Li metal anode. Electrochem. Commun. **51**, 59–63 (2015)
21. Y. Liu, D. Lin, P.Y. Yuen, K. Liu, J. Xie, R.H. Dauskardt, Y. Cui, An artificial solid electrolyte interphase with high Li-ion conductivity, mechanical strength, and flexibility for stable lithium metal anodes. Adv. Mater. **29**, 1605531 (2017)
22. Y. Lu, Z. Tu, J. Shu, L.A. Archer, Stable lithium electrodeposition in salt-reinforced electrolytes. J. Power Sources **279**, 413–418 (2015)
23. F. Wu, J. Qian, R. Chen, J. Lu, L. Li, H. Wu, J. Chen, T. Zhao, Y. Ye, K. Amine, An effective approach to protect lithium anode and improve cycle performance for Li–S batteries. ACS Appl. Mater. Interfaces **6**, 15542–15549 (2014)
24. F. Ding, W. Xu, G.L. Graff, J. Zhang, M.L. Sushko, X. Chen, Y. Shao, M.H. Engelhard, Z. Nie, J. Xiao, X. Liu, P.V. Sushko, J. Liu, J.-G. Zhang, Dendrite-free lithium deposition via self-healing electrostatic shield mechanism. J. Am. Chem. Soc. **135**, 4450–4456 (2013)
25. M. Rosso, T. Gobron, C. Brissot, J.N. Chazalviel, S. Lascaud, Onset of dendritic growth in lithium/polymer cells. J. Power Sources **97–98**, 804–806 (2001)
26. C. Brissot, M. Rosso, J.N. Chazalviel, S. Lascaud, Dendritic growth mechanisms in lithium/polymer cells. J. Power Sources **81–82**, 925–929 (1999)
27. J.N. Chazalviel, Electrochemical aspects of the generation of ramified metallic electrodeposits. Phys. Rev. A **42**, 7355–7367 (1990)
28. M. Rosso, E. Chassaing, J.N. Chazalviel, T. Gobron, Onset of current-driven concentration instabilities in thin cell electrodeposition with small inter-electrode distance. Electrochim. Acta **47**, 1267–1273 (2002)

29. H.J.S. Sand, III. On the concentration at the electrodes in a solution, with special reference to the liberation of hydrogen by electrolysis of a mixture of copper sulphate and sulphuric acid. Philos. Mag. **1**, 45–79 (1901)
30. P. Bai, J. Li, F.R. Brushett, M.Z. Bazant, Transition of lithium growth mechanisms in liquid electrolytes. Energy Environ. Sci. **9**, 3221–3229 (2016)
31. COMSOL multiphysics reference manual, version 5.3. COMSOL, Inc. www.comsol.com
32. P. Verma, P. Maire, P. Novák, A review of the features and analyses of the solid electrolyte interphase in Li-ion batteries. Electrochim. Acta **55**, 6332–6341 (2010)
33. C.-P. Yang, Y.-X. Yin, S.-F. Zhang, N.-W. Li, Y.-G. Guo, Accommodating lithium into 3D current collectors with a submicron skeleton towards long-life lithium metal anodes. Nat. Commun. **6**, 8058 (2015)
34. Q. Yun, Y.-B. He, W. Lv, Y. Zhao, B. Li, F. Kang, Q.-H. Yang, Chemical dealloying derived 3D porous current collector for Li metal anodes. Adv. Mater. **28**, 6932–6939 (2016)